AIR RAGE

Air Rage

The underestimated safety risk

ANGELA DAHLBERG
Dahlberg and Associates,
Calgary, Canada

Ashgate

Aldershot • Burlington USA • Singapore • Sydney

© Angela Dahlberg 2001

Published by
Ashgate Publishing Limited
Wey Court East
Union Road
Farnham
Surrey, GU9 7PT
England

Ashgate Publishing Company
110 Cherry Street
Suite 3-1
Burlington
VT 05401-3818
USA

Ashgate website: http://www.ashgate.com

British Library Cataloguing in Publication Data
Dahlberg, Angela
    Air rage
    1.Aeronautics, Commercial - Passenger traffic 2.Disorderly
    conduct
    I.Title
    387.7

Library of Congress Control Number: 2001087933

ISBN 978–0–7546–1325–1

Transfered to Digital Printing in 2009

MIX
Paper from responsible sources
FSC
www.fsc.org
FSC® C013985

Printed in the United Kingdom by Henry Ling Limited,
at the Dorset Press, Dorchester, DT1 1HD

Contents

List of Figures and Tables

Foreword

I had the honor and challenge of serving as Executive Chair for the landmark 1997 White House Commission-George Washington University International Conference on Aviation Safety and Security in the 21st Century. Subsequently invited to give a major presentation at an international conference on Air Rage sponsored by the Airlines Pilot Association to be held in Washington that summer, I protested that this was not an area of my expertise. I was assured there was no substantial research on this topic and that whatever psychological insights I could convey on this growing phenomenon would be welcome.

When I started to do research on air rage in preparation for the conference, I learned that what Steve Luckey of ALPA had conveyed was indeed the case, there was no systematic research on this topic. With the publication of this book, this is no longer the case. For this thoughtful, searching balanced presentation by Angela Dahlberg will in my estimation come to be regarded as the definitive work on this topic.

As I was to discover when I began to explore the topic, one of the primary reasons there was no substantial research was that there was no unified database. Each of the airlines jealously guarded its prerogatives, maintained its own records, and investigated incidents in its own unique manner, so there was no industry-wide database with comparable data to explore. Why was this? From off the record comments by frustrated cabin attendants and union officials that I interviewed, I learned there was significant pressure to not report or investigate such incidents. There were major economic concerns on the part of the industry that opening up this controversial area to public scrutiny could damage the airlines' reputations and pocketbooks.

The presentation led to numerous inquiries, among which was one that I shall always cherish, a call from Angela Dahlberg. I was swiftly to discover that in this area, which all too often has been one of media presentations of dramatic episodes, she was a unique figure, a genuine expert who had focused through her career on understanding this complex phenomenon. We quickly became friends and colleagues and team taught a new course on 'Air Rage' at the George Washington University Aviation Institute's certificate program on aviation safety and security. Unlike many so-called experts, the remarkable thing about Angela Dahlberg is that the longer I have come to know her and explore this fascinating topic with her, the greater the appreciation I have come to have for the depth and breadth of her genuine expertise.

Let me call attention to several of Dahlberg's key observations and conclusions that really reframe the problem. The question should not be, 'What is wrong with aviation passengers that they resort to such extreme behavior?'

As she trenchantly observes, the aviation passenger is but one key element in the aviation system. And it is a fundamentally different question, with the possibility of identifying different elements in the causal chain, to ask: 'What is wrong with the aviation system?' This notion of the passenger, the flight attendant, the gate attendant, the pilot, the cabin configuration, the economics and policy concerning alcohol consumption, training, and departure delays and flight cancellations in an overloaded system as integrated elements in a complicated system is of crucial importance. Dahlberg systematically analyzes these elements that she identifies as contributing to a problem with multiple linked causes, and thus there is no simple answer.

She also makes an observation that I have not seen before, which becomes self evident once she makes it—that air rage is a special case of a wide spread phenomenon in contemporary society— violence in the work place. And the approaches that have been developed in response to this problem that all too often has fatal consequences can be applied to the disruptive aviation passenger.

But the circumstances that surround the aviation passenger are unique. On entering the plane, the aviation passenger is required to totally yield control over his destiny to the crew as he straps himself into the giant metal cigar case, an inherently stressful experience. For individuals who have a strong need to be in control, the feeling of being out of control can be overwhelming. Under stress, characteristic psychological reactions become magnified. The somewhat anxious individual can become panicked. Similarly, individuals with a strong sense of entitlement, under stress can react to a mild request from a cabin attendant as if it is an outrageous affront. Consider the Saudi princess in first class who asked to fasten her seat belt, clawed at the stewardess as she screamed, 'I have never been told to do anything in my life!' Or what of the executive, used to telling his subordinates what to do, being directed to discontinue use of his laptop or cell phone?

But the stress mounts before entering the plane. The virtual epidemic of flight delays and cancellations has already led passengers to become frustrated and annoyed. And, for many, during that often long delay, alcohol consumption begins, leading to a further diminution of control over emotional reactions. Further drinking on finally entering the plane compounds the problem. Moreover, for the cabin class passenger, the increasingly restricted space between rows can lead to a feeling of personal space being invaded, and for the already claustrophobic individual, that further magnifies the stress.

Dahlberg has well dissected the complex structure if the aviation system, stressing that the sequence of events that leads to disruptive aviation passenger behavior is multiply determined. There is no magic bullet, for each of the components of that system contributes.

But how much variance each of the elements in that complex system constitutes remains unclear. One cannot fix a problem that one does not understand. A uniform system for collecting data on passenger incidents is a necessary foundation for the systematic industry wide research program that is required. In her comprehensive volume, Dahlberg points the way to the searching self-analysis of the aviation industry that is required to identify the principle elements contributing to this growing problem. This understanding in turn can lead to the necessary steps to stem the tide of the growing problem with this special case of violence in the workplace—the disruptive aviation passenger, an issue that is increasingly being recognized as a major threat to aviation safety and security.

Jerrold M. Post, M.D.[1]

Note

[1] Jerrold Post is Professor of Psychiatry, Political Psychology, and International Affairs at The George Washington University, where he is Associate Director for Safety and Security of the George Washington University Aviation Institute.

Preface

I have written this book out of a sense of empathy for the air traveler and a deep commitment to my colleagues in the aviation industry, especially the many cabin crew members who excel in the art of customer service. Both groups have taught me great lessons in conflict resolution, generosity of spirit, and the magic of kindness. Meeting and speaking with victims of assault has filled me with respect and compassion for their struggle to feel whole again. Some have taken their experiences in the open to uncover the abuse inflicted by air travelers, and have been instrumental in rallying wide-ranging support leading to change. Today, airlines and airline workers are struggling with the issue of workplace violence, as it manifests itself through passenger misconduct. Despite some significant developments over the last couple of years, mainly on the legal side, front line airline workers feel under attack.

This book is intended to be useful for aviation professionals at the operational level, in the disciplines of human factors, sociology, aviation psychology, safety and security, training and marketing. It is also intended to respond to public interest since it is the educated consumer who can play a more effective role in keeping the airline industry focused on the human side of the business.

'Air rage' is a form of workplace violence. Workplace violence in civil aviation is an under-acknowledged and under-researched issue, even today. The term workplace violence is only now gaining some acceptance by the industry. This is puzzling since the issue of workplace violence has surfaced in the mid sixties, and has touched a wide range of professions since then. There is a large body of knowledge and professional expertise resulting in a well established research and business network dealing with the prevention of workplace violence and the effective management of post incident trauma.

What are the airlines' attitudes towards any form of workplace violence, and why is it only the customer who appears to be the perpetrator? My investigation of the issue of 'air rage' revealed that airline workers rarely lodge an official complaint about sexual harassment, verbal abuse and intimidation or physical violence from their colleagues. This observation is intriguing since airline workers are not a more saintly group than any other workforce.

Part of the research for this book is based on interviews with air travelers and airline representatives from different national and international carriers. Expertise has been sought and generously provided by professionals and friends in the disciplines of psychiatry, psychology, aviation medicine, healthcare, regulatory, and activist groups. My own on-site observations of the aviation culture represent an insider view of the changing attitudes towards the air traveler, the marketing requirements affecting cabin crew[1] selection, and the impact of economic constraints on training. I have been a player in and a recipient of corporate and departmental power struggles. The ups and downs of service versus safety, the conflict and the role confusion that ensues, are part of the covert organizational culture the air traveler unwittingly enters with the purchase of a ticket. Service provider and service consumer are integral players in a service experience that starts far in advance of the actual flight.

The causes for air travelers' expressed unwillingness to comply with rules and regulations, their feeling of entitlement for quality service, and their right to express their anger when their needs have not been met, are multi-faceted. Society has changed, so have airline workers, and so have air travelers.

The phenomenon of passenger misconduct must be considered as a system issue with a full range of complexities. One of the main contributors to passenger misconduct is the aviation system itself. Driven by aviation politics, economics and production pressures, organizational routine and conformity, the aviation system has become complacent, and inwardly focused.

Evidence suggests that the aviation system has failed today's air traveler. Consequently, it has significant leverage to reduce the high percentage of incidents attributed to multiple service failures. The question is, does it have the will to do so or is the focus on legal matters a strategy to divert attention from the larger organizational and ethical issues of running the business? Expressions of conflict do not occur in isolation. The majority of cases are clearly linked to situational events and triggered by interpersonal accidents between the air traveler and the service provider.

In this book a framework for understanding the phenomenon of 'air rage' and offer solutions to reduce hostilities between the customer and service provider is provided. I devoted my career to studying the conflict between safety and service from various vantage points, at the operational level, in a training position, and at the policy-making level. I have experienced this conflict personally and developed some insight that might be helpful in achieving synergy between safety and service.

The book is divided into six chapters. The first chapter provides an overview of the evolution of the phenomenon of 'air rage'. The media played an important role in bringing the issue into the open and create public interest. In defining the issues, responses by labour, industry and regulatory authorities have focused on regulatory and legislative measures of deterrents. The fact that the fundamental issue of data collection and methodology has yet to be resolved fosters the search for external solutions when internal solutions would be more appropriate. The current emphasis on reactionary and punitive measures via legislation clouds the bigger issues. Corporate and individual responsibilities cannot be avoided.

The second chapter introduces an expanded concept of human factors. It champions the idea that the air traveler is an integral part of the aviation safety net. The air transport system has increased performance expectations of passengers due to increased security and economic needs. Affecting the basic relationship between the consumer and the airlines, the air traveler as an active participant in on-board safety has received little help to safeguard his performance level. Various models designed to assist the airline worker to focus on the essence of service and prevent hostilities resulting from conflict situations will be provided.

The third chapter presents an overview of today's air traveler, including legislative changes that have triggered increasing diversification of the airline operation. Special attention is given to travelers with visible and invisible disabilities since they are the one group that so far has never been involved in incidents of misconduct. The complex aviation system alienates the consumer while economic considerations lead to environmental risk factors within the airports and the aircraft cabin.

The fourth chapter offers a historical perspective of the role of cabin and flight crews, and attempts to identify the impact of organizational, psychological, and cultural factors on the relationship with the air traveler.

The fifth chapter covers the aviation system, organizational culture and its effects on organizational and individual performance. Examining what forces influence the adversarial relationship between the system itself and the air traveler will lead to more appropriate countermeasures. The main themes of 'safety versus service', marketing's role in influencing customer expectations, the selection of cabin crew and training are discussed here.

The sixth chapter is dedicated to passenger risk management. An interactive model that facilitates the analysis of system issues and offers a methodology for reducing risks is presented. The emphasis on information systems and current gaps is a cornerstone of this chapter leading to a model of an investigation process based on human factors criteria.

This book is written out of a deep-seated concern for social peace on-board and the safety of the men and women who face conflict as part of their professional reality. This is a time for a new vision. What is it going to be? From Nightingale of the skies, to pin-up girls, to servers to safety wardens of the skies, none of these images meet the current and future needs. The debate is just beginning.

Note

[1] The term 'cabin crew' denotes flight attendants or cabin attendants regardless of their rank, or any other term used in aviation. The term 'flight crew' denotes pilots regardless of their rank, and is consistent with the terminology used by ICAO.

Acknowledgements

What started out as a personal endeavour to investigate the phenomenon of 'air rage' back in 1995 soon became a fascination with a topic that quickly expanded in scope and importance.

My own search for better understanding prompted my involvement as an independent member of the Transport Canada Working Group on Prohibition of Interference with Crew Members. In this process, I made invaluable contacts with a diverse network of individuals who all have a stake in the issue. Unexpected friendships and professional relationships developed, encouraging me to continue to dedicate most of my time since then to explore the complexities of the issue.

Discussions with passengers ensued on many occasions while I too joined the ranks of air travelers, offering candid views on their frustrations and observations.

The support and encouragement I received strengthened my resolve to write this book. My aim was and still is, to create a better understanding of the issue by all stakeholders and improve the chances of social peace on-board.

Many people contributed to this effort, generously giving their time and expertise. They believed the cause worthy, and energized me with their confidence. I am especially grateful to these individuals and organizations:

Dr. Jonathon Aleck and Colin Torkington, ICAO
Rebecca Chute, NASA
Dr. Leonard Berry, Texas A&M University
Dr. Barry Sheehy
Marie Doll
Agnes Hibbert,Institute of Health Sciences, University of Oxford
Sylvia Loh
Dr. Helmut Jungermann, Technische Universitat Berlin
Dr. Rob Bor, London Guildhall University
Dr. Vittorio Di Martino, International Labour Organization
Reiner Kemmler, Lufthansa
Captain Matt Sheehy, Canadian Airlines International
Captain Steve Luckey, ALPA
Captain Ed Horton, American Airlines
Captain Ellis Nelson, United Airlines
Sergent Malcolm Bow, Peel Regional Police
Michael Murphy, Air Passenger Safety Group

Susan Howland, SKYHELP
Renee and Michael Sheffer, SKYRAGE
Rick Goodfellow, Independent Living Resource Centre, Calgary
The members of the Inflight Service Management Board, IATA
Transport Canada
Canadian Transportation Agency
The Aviation Institute, George Washington University
ALPA
CUPE
IFALPA
ILO

I am grateful to the media for their generous consent to include material obtained through their many channels both in print and on the Internet:
ABCNEWS, AIRWISE News, AIRJET AIRLINE WORLD NEWS, Airlinebiz, Air Travel, Airwise News, Associated Press, AVflash, BBC News World Europe, CNN News, Calgary Herald, Electronic Telegraph, Financial Post, Globe and Mail, Kuam News, Las Vegas Review Journal, Reuters, San Francisco Chronicle, San Francisco Examiner, Scotland on Sunday, Star Telegram, The Associated Press, The Edmonton Journal, The Indianapolis Star, The Seattle Times, Travel News, USA Today, and Washington Post.

Dr. Jerrold Post, one of my most critical friends, provided challenge and guidance in preparing me for publication. Robert Herron, a long time friend and associate, offered a fresh non-airline perspective and editorial assistance throughout.
As one's private life is transformed by the commitment to writing, my husband's unwavering support, affection and good humour carried me through this experience without adding any pressure. Thank you Eric.

List of Abbreviations

ACI	Airport Council International
ACPA	Air Canada Pilots Association
ADA	Americans With Disabilities Act
AED	Automatic external defibrillator
AFA	Association of Flight Attendants
AIR	Aviation Investment and Reform Act
ALPA	Airline Pilot Association International
ALPA-SA	Airline Pilot Association-South Africa
APFA	Association of Professional Flight Attendants
APSG	Air Passenger Safety Group
APU	Aircraft power units
AQR	Airline Quality Rating
ASHRAE	American Society of Heating, Refrigeration and Air Conditioning Engineers
ASRS	Aviation Safety Reporting System ATA Air Transport Association of America
ATB	Automated ticketing and boarding
ATC	Air Traffic Control
AVOSS	Aircraft Vortex Spacing System
BAT	Boeing Air Transport
CAA	Civil Aviation Authorities
CAB	Civil Aeronautics Board
CALFAA	Canadian Airlines Flight Attendant Association
CASE	Civil Aviation Security Enhancement
CRM	Crew Resource Management
CTA	Canadian Transportation Agency
CUPE	Airline Division of the

	Canadian Union of PublicEmployees
DOT	Department of Transportation
ET	Electronic ticketing
FAA	Federal Aviation Administration
FBI	Federal Bureau of Investigation
IASA	International Aviation Security Academy
IATA	International Air Transport Association
IAM	International Association of Machinists
IASA	International Aviation Security Academy
ICAO	International Civil Aviation Organization, role 159
ICVS	International Crime Victim Survey
IFALPA	International Federation of Pilot Associations
IFFA	Independent Federation of Flight Attendants
ILO	International Labour Organization
ILRCC	Independent Living Resource Centre Calgary
ITF	International Transport Workers' Federation
NASA	National Aeronautics and Space Administration
NATA	National Air Transportation Association
NTSB	National Transportation Safety Board
OIG	Office of Inspector General
PACE	Passenger-Centric Equipment
PRP	Peel Regional Police
PSA	Pacific Southwest Airlines
RCMP	Royal Canadian Mounted Police

STOC	Station Operations Control Centre	TWU	International Brotherhood of Teamsters and Transport Workers Union of America
TCA	Trans-Canada Air Lines		
TSB	Canadian Transportation Safety Board		

1 Air Rage – Developing a New Understanding

Background

The issue of passenger misconduct, ranging from verbal abuse to assault in the aircraft cabin, has achieved wide spread recognition in recent years as a result of victims' advocacy, labour and industry initiatives, and media attention. The term 'air rage' is a label now used by the media not only for high profile cases but encompasses all forms of passenger behaviour causing a disturbance. Not formally defined within the airline industry or in academic circles, the airline industry prefers to use euphemisms such as disruptive and/or unruly passengers when referring to on board incidents of passenger misconduct. Crew member labour associations, on the other hand, refer to such incidents as interference with crew members. The term 'interference with crew members' stems from existing legislation and defines actions that disrupt crews in the performance of their job functions. The choice of terminology is significant, each pointing to different interests vested in the occurrence of passenger misconduct. The media looks for sensational headlines; airlines are more cautious and aim to strike a cautious balance, while crew member labour associations immediately imply the vilification of the air traveler, with punishment as a main source for retribution. The Transport Canada Working Group formed in 1998 saw the different stakeholder interests played out in its first round of meetings.[1] Initially called 'Working Group on Abusive and Unruly Passengers,' representatives of the Canadian Union of Public Employees (CUPE) recommended changing the title of the Working Group to 'The Working Group on Interference with Crew Members'.[2]

This book employs the term passenger misconduct in most discussions. It allows for improved understanding of the wide range of its manifestations, and provides the opportunity to position the issue in the context of the more global issue of workplace violence.

Definitions

The following definitions have been developed recognizing a broad range of definitions used by professionals worldwide in the context of workplace violence. These definitions differ from those used by various jurisdictions.

- ***Violence***: Any incident in which an employee is abused, threatened or assaulted by a co-worker or member of the traveling public in situations connected with his or her employment.

- ***Verbal Abuse and Threats***: Physical contact does not occur. Incidents include rude gestures, innuendoes, sexual and racial harassment, harassment because of a persons sexual orientation, intimidation to report you, to have you fired, being yelled at, cursed at.

- ***Assault***: Physical contact does occur and results in either no injury or injury. Incidents include being grabbed, pushed, shoved, hit, punched, thrown, etc.

- ***Sexual Harassment***: Incidents include unwanted conduct of a sexual nature, or other conduct based on sex affecting the dignity of men and women at work. Therefore, a range of behaviour may be considered to constitute sexual harassment. Unwanted behaviour may include suggestive gestures, sexual innuendoes, advances, etc.

- ***Sexual Assault***: Incidents include physical contact of sexual nature, groping, fondling, rape.

Gaining Attention

Today's worldwide coverage of incidents by the media has expanded both on television, and in the printed press leading to a growing interest by the public, industry and regulatory authorities, including academics to address this issue. The Internet is a prolific source for material on this topic, contributing to a greater understanding of the issues involved while also offering the public a venue to express their opinion.

Actions of misconduct and assault are predominantly aimed at crew members however fellow air travelers, be they complete strangers, friends or family members are not spared. Increased concern and awareness by the public have given support to strategic efforts by national and international flight and cabin crew member labour associations to obtain cooperation from airlines and civil aviation authorities to deal with this phenomenon, and address the wider range of legislative and regulatory issues.

In 1995, C. Laurence, an expert in trauma and disaster psychology, positioned passenger misconduct in the context of workplace violence. In her paper titled *Violence in the Workplace/Airplane Cabin*. Laurence sees North American culture with its propensity for violence as the societal landscape for on board display of conflict. According to Laurence, Maslow's hierarchy of needs allows us to understand passenger aggression, as it relates to the deprivation of those needs. Maslow introduced the concept of five levels of human needs. The lowest need is physiological such as hunger, and thirst, the second level of need is safety, for example security, followed by the need to belong, affection, and identification. The fourth level deals with esteem needs, for example, prestige, and self-respect, and the fifth level deals with the need for self-actualization.

Laurence presented the problem in very human terms with a focus on the air traveler as a person with certain needs the airlines have a responsibility to address. Her recommendations ranged from skill enhancement to conflict resolution training, to reporting of such incidents, post traumatic counseling and support to the adoption of a corporate zero-tolerance policy. Clearly, the airlines were seen to be in a key position to address causes for on-board conflict. Since then, consumers, researchers, academics and practitioners also link service failure and a lack of empathy and caring staff amongst other factors to incidents of passenger misconduct.[3]

Forms of Passenger Misconduct

Passenger misconduct in the form of bullying airline representatives on the ground or in flight is not new. Senior cabin crew members will easily recall personal experiences going back to the late 60s. The introduction of jet aircraft resulted in lower fares and the beginning of mass air travel. The new era of technology in airline operations in general changed the social make up of the air traveler. So did the relationship between the airlines and their customers.[4]

Incidents of bullying are usually related to a situation of real or imagined service failure when passengers feel justified in verbally attacking the cabin crew, undermining their official authority, personal dignity and professional competence. Examples are wide ranging and include blaming cabin crew members for the lack of amenities, meals, delays, challenging their authority when requested to comply with safety rules, and making personal attacks, be they verbal, or physical

Sexual harassment often delivered verbally, but also including unwanted touching or groping was something cabin crew members experienced as part of the working conditions The cabin crew image was reduced to a sex symbol when the airlines followed a fashion trend and introduced mini skirts and dresses as a major component of cabin crew uniforms. The suggestive airline advertising at that time, as exemplified by the provocative stewardess, with the ad captioned, 'I'm American Airlines – fly me' suggested the sexual availability of the cabin crew members.

Sexual escapades by members of the 'mile high club', who actively seek the thrill of unusual places to enhance their performance, are another type of misconduct. Others may just get carried away with total strangers, encouraged by alcohol and the close vicinity of an attractive stranger. Considering these activities as potential interference with crew members, the Transport Canada Working Group concluded that it was not so. Since then however, one such case resulted in prosecution.[5]

Cases of physical assault were very rare and were usually handled through the intervention by the captain with no report submitted to airline management. The reason for the lack of reporting is rooted in the airlines' service culture that pronounced 'the customer is always right', but was also a reflection of the airlines wishing to avoid unseemly publicity. In this environment, the stewardess who complained was-and is- often treated as a trouble masker, leading to a blanket of silence.

Not the Only Victims

The job of the cabin crew offers parallels with other jobs and professions in terms of job content, task situations and the outcome of such incidents. The findings of the International Crime Victim Survey (ICVS),[6] considered to be the most comprehensive effort of standardized sample surveys, focus on householder's experience with crime, policing, crime prevention and feelings of unsafety, provide useful insight on a number of issues. Data spanning the time period from 1989 to 1997 and involving more than 50 countries, tracks the trends in crime.[7] Of particular interest is that violence is a common occurrence when workers are in contact with people in distress. Chappell and DiMartino (1998) explain in 'Violence At Work':

> Frustration and anger[8] arising out of illness and pain, old-age problems, psychiatric disorders, alcohol and substance abuse can affect behaviour and make people verbally or physically violent. Increasing poverty and marginalization in the community in which the aggressor lives; inadequacies in the environment where care activities are performed, or in the way these are organized; insufficient training and interpersonal skills of staff providing services to this population; and a general climate of stress and insecurity at the workplace can all contribute substantially to an increase in the level of violence.

It follows that caregivers are amongst the high-risk workers, together with hotel, catering and restaurant staff. Similarly, airline workers who deal with the public face passenger frustration and anger arising out of a general climate of stress and insecurity as well as service problems. The airline worker has the disadvantage of not knowing what individual risk factors, such as psychiatric disorders, alcohol and substance abuse afflict a passenger and not being trained to deal with such cases, in contrast to a caregiver in the mental health profession.

Naming and Blaming

In September 1996, TV journalist Barbara Walters featured a segment on abusive passengers, and Oprah Winfrey dedicated a full hour to the 'ugly passenger later that same year. Ms. Anne McNamara, American Airlines Senior Vice President, appearing on CBS's 20/20 announced the company's implementation of a non-tolerance policy and stated: '... we weren't as supportive (to our crewmembers) in the past as we are today. We were perhaps caught up in the theory that the customers are right, and we finally decided that there are some customers we just don't want.' This statement by an airline executive may well be the first widely broadcast admission that the operating values of the aviation industry and its organizational culture are in need of change.

Alcohol consumption, drug dependencies, mental instability, gambling losses, special charter groups and sports teams, seasonal workers, and other types of group travel, as well as certain operational situations such as delays or aircraft diversions, all appear to be conditions that may lead to disturbances in flight.[9] Despite these more obvious factors, we have little understanding of why some people lose control when most passengers are able to cope with the more unpleasant aspects of air travel. The tendency is to look for single causes and simple solutions. Apportioning blame does not solve the problem.

Links to Flight Safety

Reports filed on a confidential basis with the National Aeronautics and Space Administration's (NASA) Aviation Safety Reporting System (ASRS) provide some figures on the number of flight crew errors as a direct result of their involvement in passenger misconduct incidents. I conducted an analysis and presented the findings at the Tenth International Symposium on Aviation Psychology in 1999. A total of 19 safety infractions or 21 percent of the 91 reported since January 1, 1988 provide some important details. Flight crew errors ranged from altimeter deviations, overshooting destination, unauthorized runway entry, non-adherence to donning oxygen mask, speed deviation during descent to less than legal separation to flying off course to landing with unsecured cabin.

In 19 of the 91 incidents the captain or his co-pilot left the cockpit to personally intervene in a passenger misconduct situation. Surprisingly only 4 of these incidents are connected with a resulting operational performance failure. This finding is interesting since flight crew unions and labour associations make a strong case for the need to remain in the cockpit at all times. It appears that the risk of flight crew error increases four times when a member of the cabin crew enters the cockpit to communicate in person with the pilots on matters related to passenger misconduct. This preliminary finding identifies a need to further analyze such incidents with the view to reassessing current communication procedures between the flight crews and the cabin crews.

A subsequent study released by NASA (Connell, Mellone, & Morrison, 2000), covering ASRS passenger behaviour for the period of January 1999 through December 1999, focuses on 152 incidents reported in that category. Sixty of these incidents resulted in flight crew error while in almost a quarter of these incidents a flight crew member left the cockpit to investigate the situation. The analysis falls short in that it does not examine the relationship between crew errors incurred while the cabin crew reported on a passenger incident in the cockpit compared to crew error incurred when one of the flight crew elected to leave the cockpit to be briefed by the cabin crew.

These studies point to the need for further research examining the pros and cons of flight- and cabin crew communication, where and how it should be done. Determining what communication protocol would achieve a reduction of pilot error is the main concern since the absence of such a protocol creates a range of risk factors for the safe operation of the aircraft. The significance of passenger misconduct on the performance of all crew members is not in dispute. Crews are trained to deal with operational irregularities and emergencies in a very structured way. This is not the case when confronted with highly emotionally charged situations in the passenger cabin.

Strategies of Change

The primary focus on 'crime and punishment' is not surprising given the severity of some of the physical assaults suffered by cabin crew members and reported by the media as early as in 1995. Deterrents are one of the many strategies needed to address this problem. Some of the stories were reported worldwide and have resulted in a concerted effort by a group of cabin crew labour associations to formally address the issue in December 1995. Union officials from the Airline Division of the Canadian Union of Public Employees (CUPE), Association of Flight Attendants (AFA), Association of Professional Flight Attendants (APFA), Independent Federation of Flight Attendants, International Association of Machinists (IAM), International Brotherhood of Teamsters and Transport Workers Union of America (TWU), formed the Coalition of Flight Attendant Union's Assault Task Force.

In its early stages of examining the difficulties of dealing effectively with passenger misconduct, the Task Force determined the need for enforcement guidelines, a common policy for airlines to specifically address passenger misconduct and assault, and education for the air traveler.

The phenomenon of violence in the aircraft passenger cabin knows no international boundaries. It affects air travel worldwide, although to what degree is only slowly emerging. The Asia-Pacific Cabin Safety Working Group, originated in 1993, represents Australia, French Polynesia, Hong Kong, Japan, New Zealand, Papua New Guinea, Singapore, Solomon Islands, South Africa, Taiwan, and Western Samoa. One of its sub-groups is specifically dealing with the topic of passenger misconduct.

Punishment of the perpetrator continues to be foremost on the agenda in 1999 as evidenced by the initiatives taken by countries such as the United States of America, Canada, and the United Kingdom. The International Civil Aviation Organization (ICAO) initiated a working group to deal with the global aspects of security and legal issues surrounding passenger misconduct. Intense lobbying by the International Air Transport Association (IATA), the International Federation of Pilot's Associations (IFALPA), representing in excess of 120,000 airline pilots in almost one hundred countries, the International Transport Workers' Federation (ITF) and air passenger associations, preempted these activities.[10]

Reliability of Data

The study of passenger misconduct is faced with a number of problems. The number and severity of incidents appears to be on a steady rise, but there is a wide disparity among the diverse sources reporting. Figures cited by flight and cabin crew labour associations differ from figures cited by industry, IATA, ICAO, and NASA. Moreover, there are striking differences among reports by air carriers. These figures are problematic for comparative purposes because of the differences in legal definitions and reveal the need for standardization in the areas of precise rules for classifying incidents, recording practices, and counting.

Loh, in her thesis 'An Investigation of Cabin Safety: The Taxonomy of Disruptive Passenger Behaviour' (Swinburne University, 2000), developed a theoretical framework for effective data collection and analysis. This system is designed to classify and code information more easily to assess the causal factors contributing to passenger misconduct. To date, her approach offers the first major advance in offering a methodology that would facilitate the development of improved prevention strategies.

Since the IATA Inflight Service Management Council placed the issue of passenger misconduct on the agenda in 1996, airlines implementing a zero-tolerance policy, saw their reports on passenger misconduct escalate significantly. This rise was as a direct result of a fundamental culture change in these organizations. Cabin crew members did not previously report passenger misconduct since they knew that their organizations did not provide any support.

The steady increase of reports is subsequently a combination of cabin crew members feeling encouraged to report such incidents, law enforcement also focusing their attention on the issue and keeping better records, and media fascination with the more extreme cases. The problem of an unreliable database persists however, for a global definition of passenger misconduct, abuse and violence, and a global methodology for collecting such data, still does not exist. This is not unusual and mirrors the same problem reported by Chappell and Di Martino in 'Violence at Work', a 1998 publication of the International Labour Organization in Geneva. A number of steps taken by aviation organizations in 1999 show some progress towards developing a methodology to examine passenger misconduct.

On the Agenda

Different levels within the aviation system addressed the issue of data gathering on incidents of passenger misconduct. The Air Transport Association of America (ATA) Cabin Operations Committee for example has developed a standardized database for a voluntary reporting program for member airlines. The objective of this database is twofold; to produce a baseline for incidents related to passenger misconduct and to analyze trends. This data could also be used to measure the impact of initiatives that either prevent or diffuse disruptive situations or identify key problems surrounding these events. This industry database became operational in April 1999. The first report covers the period of April, May and June 1999.

The Transport Canada Working Group On Prohibition Against Interference With Crew Members raised the issue during its early deliberations.[11] As The Group examined the pro and cons for maintaining government statistics on incidents of passenger abuse and violence, the majority of the members agreed:

> ... that there should be a regulatory requirement for air carriers to establish an incident reporting system and that reporting of incidents by employees should be voluntary. The regulation should provide parameters for information requirements, based on a model incident reporting form. selected data gathering achieves little in scooping the issue objectively.[12]

The airline industry has major concerns about regulatory requirements for collecting data on passenger misconduct. Issues of liability, confidentiality and corporate image are at stake. Although there is agreement by the members of the *Transport Canada Working Group On Prohibition Against Interference With Crew Members* to develop and implement a database for Canadian carriers, the quality of the statistics, which would be made available to the consumer, is not guaranteed. Legal questions concerning the workers' right to know about workplace hazards were discussed, however what information the consumer should have prior to making a decision to purchase a seat from one carrier or the other, was not explored.

Following suit, the United Kingdom Civil Aviation Authority formed a Working Group in 1999 to collect industry data. As helpful as these initiatives may seem, little use is evidently made of research material already available in the area of workplace violence, and in providing an understanding of what information needs to be collected for a long term approach to deal with this phenomenon. The design of these reporting forms varies in their number of data fields and elements and often focuses on a limited number of situations. As a result these databases do not:

- Support a move towards global comparability;
- Assist in creating a more thorough understanding of the interaction of different factors culminating in passenger misconduct;
- Offer a basis to fully explore preventive measures.

Legal Framework

Secondly, issues surrounding the complex legal framework are being taken into account, both in the national and international arena. This includes enforcement issues. Formal discussions at various levels within the aviation industry resulted in the International Civil Aviation Organization (ICAO) to put the issue of unruly passengers on its agenda. Non-governmental organizations participate in ICAO's work and include the International Air Transport Association, the Airports Council International, the International Federation of Air Line Pilots' Associations, and the International Council of Aircraft Owner and Pilot Associations.

At the meeting held April 29 - May 9, 1997, the ICAO Legal Committee gave the matter of abusive and violent passengers some attention by concluding this issue should be studied and the Council should assign a priority to this issue. The first meeting of the ICAO Secretariat Study Group On Unruly Passengers took place in Montreal, on January 25 to 26, 1999. This study group is formed under the auspices of the Legal Bureau, one of the five bureaus of the ICAO Secretariat.

The one hundred and eighty-five Contracting States of ICAO present a challenging variety of many legal philosophies and many different systems of jurisprudence. As evidenced in the case of passenger misconduct, there is need for the development of a unified code of international air law. It is within ICAO's mandate to facilitate the adoption of international air law instruments and to promote their general acceptance.

The Assembly, composed of representatives from all Contracting States, is the sovereign body of ICAO. The Assembly meets every three years, reviewing in detail the work of the Organization and setting policy for the coming years. The Council, the governing body, gives continuing direction to the work of ICAO. It is in the Council that Standards and Recommended Practices are adopted and incorporated as Annexes to the Convention on International Civil Aviation. ICAO works in close co-operation with other members of the United Nations family.

During the briefing of the Air Navigation Commission on The Subject of Unruly Passengers in Montreal in May 1999, aviation security experts presented their assessments and recommendations to combat incidents of passenger misconduct onboard. The recommendations included:[13]

- A unified code of international air law;
- Educational campaign for the public;
- Standard requirement for lockable cockpit door;
- Flight crew not to leave cockpit to intervene in passenger misconduct incidents;
- Training for ground workers interacting with the public and crews;
- Training cabin crews in restraining methods;
- Standard Training Package (ICAO);
- Critical Incident Stress Programs;
- Consideration of the 'jump seat rider' as a resource;
- Incident reporting forms;
- 'Notification to Passenger' forms;
- Preventive measures to include the entire process from the arrival at the airport to the departure from the airport.

Corporate Culture

The airline's corporate climate and culture create a work environment that affects both group and individual performances. This connection has been the subject of research for some time, obtaining special recognition at the first Corporate Culture and Transportation Safety Symposium, sponsored by the National Transportation Safety Board (NTSB) in April 1997. (Ginnett, 1997; Hayward, 1995; Marske, 1997; Reason, 1997; Suggs, 1997; Westrum, 1995).

With mergers, labour unrest, ever-increasing workload and changing technology, human error is more likely to occur. Between system designs, worker stress, poorly designed safety policies and inadequate support systems, there is much concern for operators of the technical systems.

Important aspects in developing an understanding of passenger misconduct are organizational issues such as:

- Organizational culture;
- Marketing;
- Recruitment and training;

- Customer service policies;
- Operational policies;
- Communications;
- Performance standards and measurements.

Frequently the very components of an organization are at war with one another with such conflicts carrying over to the workers interacting with the public, and the public itself. Typically of the scope of aviation research to date, the effects on the issue of passenger misconduct has not been explored.

Service Quality Issues

Service quality, especially as measured by courteous and caring interaction with airline staff, prompt and efficient communication when things go wrong, indicate a steady decline in the eyes of the air traveler, especially in the U.S. The airline industry expects a certain standard of passenger behaviour that conforms to the processes designed for efficiency. These processes, predominantly designed for economic efficiency, show little consideration for human social conventions, habits and personal space requirements. The broad accessibility of air travel to all levels of society requires greater sensitivity to passengers' need for high-touch service in a high-tech environment. The demands of the socio-economic background of people, and psychological factors involved associated with air travel need to be acknowledged and integrated in both service design and delivery to minimize incidents of passenger misconduct.

Service failures experienced prior to boarding the aircraft can be antecedents to overt conflict in the passenger cabin when cabin crews cannot meet a passenger's need immediately because of safety tasks having priority over service tasks during critical phases of the operation. Airline service policies and procedures and marketing strategies have also been identified as potential to have a negative effect on the prevention of passenger abuse and violence. Evidence and detailed coverage of these factors is provided in Chapter 5, 'The Aviation System'.

Considering service quality strategies, research has demonstrated their many benefits on customer relations and profit margins. No doubt, the existing conflict of safety versus service is creating continuous tension in how cabin crews perceive their own role onboard and confusion for the air traveler.

Airlines are faced with the formidable task to persuade their workers of providing quality service to their customers. Management clichés are not enough to make this happen. Research on a human factors approach offers some insight and practical consideration. After all the drifting, a new vision of safety-service synergy is a straight line out of the controversy.

Airport and Aircraft Cabin Environment

Airport design and processes impose behavior restrictions on passengers, leaving many with a sense of personal impotence and heightened stress. Repeatedly forced to line-up when entering a new phase in the overall passenger process, air travelers become easily frustrated and irritable. Crowded airports, especially during inclement weather cause stranded passengers to clamber for attention and resources, triggering aggressive behaviour in some. Regulatory requirements for security and safety and changes in non-smoking legislation add to restrictions of passenger freedom and likely play a part in explosive behavior outbursts.

IATA in cooperation with the Airport Council International (ACI) addressed some of these issues during a conference in October 2000, addressing security problems caused by acts of unlawful interference by passengers.[14]

Over the last four decades, the link between the unique aircraft cabin environmental conditions and aggressive behaviour has received the attention of isolated research.[15] No doubt further research initiatives will mushroom with the surge of attention given to passenger misconduct.

Issue Ownership

The five main interest groups with a primary stake in this issue are the professional flight and cabin crew member labour associations, the victim, the airlines, aviation regulatory authorities and the consumer. The media focus on the victims' stories has had a significant impact on the developments and the chronology of events affecting much needed changes to curb the seeming escalation of passenger misconduct. Security experts, the legal community and enforcement responded by rallying to work on a more effective penal system. The potential hazard to the safe operation of the aircraft because of passenger misconduct has been the main argument, although incidents involving interference with flight crews have been few.

The issue of passenger abuse and violence is not new. What *is* new is the airlines' attitudes towards this reality, its recognition that changes must be made because of legislative requirements, liabilities towards its workforce and the consumer. Furthermore the reported frequency and severity of incidents appears to have escalated to a high level, demanding legislative changes as well as cultural changes within airline organizations. Airlines are one of the last major global employers to face this issue head on. Other industries and professions have addressed the problem for years. Risk identification, prevention and post incident management are some of the major advances in these industries. The knowledge gained can contribute to a better understanding of the causes and remedies for aviation passenger misconduct.

Scope

Despite the fascination by the media and the public, funding for research on passenger misconduct has not been forthcoming. Neither labour associations nor industry associations or airlines have felt compelled to do so. There are a number of reasons for this: lack of commonly agreed to methodology to develop a meaningful database, diverging interests of the various organizations, legal considerations, airline concerns with their reputation, and economics associated with driving safety priorities.

One of the exceptions is a survey of the world's airlines conducted in 1999 (Bor, Russel, Parker, and Papadopoulos).[16] It focuses on four general questions aimed at record keeping, training, guidelines for crew, and shared views on the possible causes of air rage.

Current statistics are limited and fail to provide a broader understanding. They offer glimpses into the phenomenon rather than a more comprehensive look at the multiple factors triggering passenger misconduct. For this reason, insights will be sought outside of aviation research and relevant research from other disciplines will be used.

Results from consumer research such as the *Air Travel Consumer Report,* issued monthly by the U.S. Department of Transport, have riveted media attention on the decline and consequences of an over-burdened aviation system on Industry performance. The report tracks performance in four main areas:

- Flight delays;
- Mishandled baggage;
- Over-sales;
- Consumer complaints.

The findings reveal the impact of declining service quality on rising level of consumer discontent. Consumer dissatisfaction is a breeding ground for passenger misconduct, although it does not give an individual the right to act in violation of civility and the law.

The International Labour Organization (ILO) has produced important material by providing an overview and insight into workplace violence on a global scale. This work serves to launch our investigation and points to the realization that aviation professionals working with the public are very much part of the broader context of general workplace violence. Their experiences are shared by thousands of workers in other settings. Parallels exist with health care workers, and with land and marine transportation involving the carriage of passengers. These industries provide a basis for identifying both differences and common issues. Some examples of innovative changes to the workplace in other industries may prompt more speedy advancement and thoughts for the improved design of the aircraft interior.

What is different is the workplace itself. It is a workplace in motion, operated by human beings, and crosses borders and nations. As a means of transportation requiring highly specialized skills to operate in the highly sophisticated and technically mature civil aviation system, it is also vulnerable to interference from passengers unwilling or unable to follow basic safety regulations.

Passenger misconduct and violence so far has culminated in non-fatal assaults on the victims rather than in homicides. The death of two alleged perpetrators following restraint by crews with the assistance of fellow passengers points to a new problem.[17] [18] Although flight and cabin crew associations have repeatedly warned of the limits to their ability to secure the safe operation of the aircraft, the death of a perpetrator is a very high price to pay. The airlines have a significant stake in this issue. Concerns range from public image to liability to cost to obligations under the law to protect their passengers and workers from harm.

Summary

Passenger misconduct is not new. In our view, the phenomenon of passenger misconduct, abuse and violence is a system issue with a full range of complexities. It takes place in a mobile and unique physical environment with its own set of operational risks. The phenomenon of air rage is another symptom of the larger problem of global workplace violence.

Most, but not all, incidents occur as a result of some form of interaction between the airline employee and the passenger. The majority of these cases indicate a degree of non-compliance with rules and regulations. In rare cases, passenger misconduct manifests itself without the perceived provocation of a request for compliance. Mental instability, drug abuse or a combination of psychological and behavioural disabilities appears to be at the root of these instances. Adding to this is the cumulative effect of stressful situations imposed by the current aviation system. Reports on consumer complaints provide ample evidence to this effect.

Airlines' zero tolerance policies and support to its workers is fundamental in obtaining reports leading to a more comprehensive measurement of the scope of the problem. Alleviating fears of reprisal and providing assistance in the form of post incident counseling and advice concerning legal matters are at the core of zero tolerance policies.

Likewise, ambiguities concerning crew communication need to be further examined with a view to develop clear procedures and guidelines to reduce the risk of miscommunication and flight crew errors as reported in the example of the ASRS analysis. It is questionable if this can be accomplished without regulatory intervention to set the context in which these aspects should be addressed.

International legal jurisdictions and enforcement capabilities have a bearing on the issue. These are being addressed with further advancements reported at various levels in Chapter 6.

Policies and procedures, service design, selection criteria, and training are still lacking the attention needed for achieving a balanced and multifaceted approach by the airlines themselves. Education of the airline community on the issue of passenger misconduct is increasing, although critics argue that the focus is weighted in favour of workers' legal rights.

The assessment of risk to both airline workers and the traveling public is still under review. High-profile incidents reported in the media serve the public's need for instant information while the quality of the information becomes secondary to fulfilling this need. The informative nature of the media does not necessarily equate with public education.

Public education is primarily a shared responsibility between the airline industry and regulatory authorities. The media and the traveling public are dependent on these sources for factual information that will reduce biases, fear of flying and unwarranted hysteria. Much needs to be done in this area.

Making this information easily accessible is a basic requirement for providing full knowledge of regulatory restrictions, medical guidelines for those suffering from certain conditions, onboard services, compensation policies, and airline liability. Timely and accurate information in planning a trip will mitigate air traveler expectations while effective communication with customers when things go wrong, go a long way in achieving social peace in the passenger cabin. The distribution method is equally key: techno savvy individuals prefer the freedom of electronic access while individuals unfamiliar with its use are more dependent on face-to-face communication.

Notes

[1] This issue was raised during the first meeting of the Working Group on 25 May 1998.

[2] The recommendations were adopted: Meeting Summary on October 6-8, 1998.

[3] Currently there is no system model available that would assist in measuring the sequence of events and their impact on the final outcome, passenger misconduct.

[4] Bor (1999) offers three explanations for the origins of on-board violence: 1) It is an entirely new problem emerging in the mid-1990's; 2) co-factor theory; 3) epidemic curve theory. According to my investigation the epidemic curve theory is the most applicable.

[5] See chapter 6, section on 'Successful Prosecutions.'

[6] Sponsored by the Dutch Ministry of Justice; Working Group formed in 1987; 1st survey 1989; 2nd 1992; 3rd 1996/97.

[7] Sample data available on the Internet at: leidenuniv.nl/group/jfer/www/icvs.

[8] Studying the thought processes of angry and violent people, psychologists have determined that profound and often distorted perceptions of having been wronged are at the core of rage. Grievances may or may not be based in reality with small slights constituting threats, and leading to resentment and in extreme cases to violent acts.

[9] Baum, Singer, and Baum (1981) developed the 'Environmental Stress Model' offering a framework for the relationship between the human environment and human behaviour.

[10] Briefing of the Air Navigation Commission on the Subject of Unruly Passengers, Montreal, Canada, 28 May 1999.

[11] Information requirements, including incident reporting: Meeting Summary, 25-26 May 1998.

[12] Incident Reporting, Meeting Summary, 1-2 June 1999.

[13] Captain P. Reiss, 'Briefing of the Air Navigation Commission on the Subject of Unruly Passengers', 28 May 1999.

[14] IATA Symposium 'Securing The Future', October 17–20, 2000. Papers presented by Mel Littler, Manchester Airport titled 'Creating Passenger Friendly Airport Security', and Ian Jack, British Airways 'Unruly Passenger Update', specifically addressed the prevention and status of passenger misconduct.

[15] Beh and McLaughlin (1991), Denison, Ledwith and Poulton (1966), Edwards and Edwards, (1990), Lucas (1987) and Rayman (1997) have demonstrated a link between aggressive behaviour and the cabin pressure, low humidity, physical confinement and noise level.

[16] Survey results are published in the April 2001 issue of the International Civil Aviation Organization Journal.

[17] BBC News World Europe, 5 December 1998: The incident involved a man traveling on a Hungarian airline who died after having been restrained and given a sedative after becoming violent.

[18] *Air Crash Rescue News*, 17 September 2000 & *Airwise News*, 18 September 2000. The autopsy apparently revealed that the young man who tried to break into the flight deck on a Southwest Airlines flight, was strangled. Bruises and scratches were found on his torso, face and neck.

2 The Air Traveler – a Human Factors Issue

Introduction

Frantz Fanon, French psychiatrist and revolutionary writer whose work profoundly influenced the radical movement in the 1960's in the U.S. and Europe, defines violence as a cleansing force: 'It frees us from our inferiority complexes, it makes us fearless and restores our self-respect.'

If this is so, why does violence occur in a system designed to provide 'safety and comfort'? What are the contributing factors, and what can be done to achieve social harmony on board an aircraft? These questions need to be explored in order to design preventive measures to curb the unsettling trend of rising passenger misconduct. Until recently the emphasis has been placed on reactionary measures, from legal and law enforcement responses to training airline personnel on measures of restraint and procedures following an incident.

Although this development plays an important part in formulating a multifaceted strategy to address the problem of passenger misconduct, it also strains the uneasy relationship between the consumer and the airlines. Airlines, rightly or wrongly, are seen as ruthlessly pursuing profits without due regard for the consumer. Because air travelers experience a growing disregard for their needs, the industry imperils its prime mission to provide safe transportation.

Passenger misconduct incidents have already proven to contribute to operational system errors. Cabin crew member testimonials continue to give evidence to the effects of such incidents on their lives.[1] Clearly, the focus is on the air traveler as a potential perpetrator thereby widening the rift between industry and the consumer. The air traveler is simply another human factor in a system that has grown very fast and places a high degree of trust in automation to solve all its problems. As consumer complaints reveal, the aviation system no longer delivers on its service promise.

This chapter champions the idea that the air traveler is an under-rated human factor in the aviation safety network (Dahlberg, 1997a, b, c).[2] Although developments on the regulatory and technology front have advanced efforts for safer air travel, the air traveler as an active participant in on-board safety matters has received little positive acknowledgement.[3] I will examine the relationship of the aviation system with the air traveler and the increased expectations of standard passenger performance. In support of a preventive

17

strategy under direct control of airline management, the discussion focuses on service approaches rather than punitive or reactive approaches. These concepts have been effectively used in our own work (DAHLBERG & ASSOCIATES) within the industry and served to bridge the conflict between safety and service. How do you define service? Service quality expert Leonard L. Berry at Texas AM University explains:

> ... service is a blending of server and customer performances, equipment, materials, and facilities. Many opportunities for glitches exist.

Airlines generally excel at blending technology with the performance of their workers. Looking at service in this way, it is clear what has been missing most, the blending of these components with server and customer performances. Airline marketers thrive on blending facilities, materials and equipment in the quest for customer loyalty. Helping airline workers to understand the customer performance issues is the focus here. Both sides need a clearer understanding of their roles, and give fresh cohesion to their relationship.

Human Factors in Aviation

Human Factors is the study of how people interact with their technological environment. The focus of human factors' research in aviation is directed towards the operators of the aviation system itself. They, together with the regulatory authorities, assume responsibility for providing safe transportation to the air traveler worldwide.

Since Orville Wright accomplished the first flight on Thursday, December 17, 1903, the focus on the human-machine interface prompted by the need for safety has resulted in prolific research with the aim to improve both human and mechanical performance. Over nearly a century of human factors' research, significant changes and expansion of research interest have given much insight into the capabilities of workers in the aviation system and their relationship with technology. Human factors' research and development has exploded along with technical advances, especially computerization and industry's need for economy of operation. Aircraft accident investigations have also played a significant role in the identification of human factor issues in the cockpit, with air traffic control, and communication with crewmembers. Major accidents have not been in vain. Recommendations by safety boards and inquiries have led to significant changes over time, benefiting aviation safety to a very high degree.

Redefining Human Factors

The emergence of 'air rage', touted as a new phenomenon by early media reports of shockingly abusive and violent passenger behaviour, raises new questions concerning the need to expand the traditional definition of human factors to include the end user of the system, the air traveler. The reasons for such consideration are many. An analysis of reports filed on a confidential basis with NASA's Aviation Safety Reporting System (ASRS) confirm a number of pilot errors as a direct result of their involvement in a passenger misconduct incident. Fortunately, these errors or performance failures have not resulted in an aircraft accident so far. They do however send a clear signal of alarm, warranting much closer investigation with the aim to design preventive measures both for cockpit and passenger cabin management. The focus here is on the passenger as an under-appreciated variable in the overall effort to maintain a safe flight operation, not just a commercial commodity. In this context I redefine Human Factors as the study to include:

> ... how air travelers interact with the aviation environment; and how the performance of the air traveler is influenced by such elements as airport design, noise level, signage, automation and information overload. Human Factors examines how aircraft design, seating, temperature, vibration and oxygen, the influences of policies and procedures, service design and products affect passengers' bodies, emotions and performances in their interaction with each other and airline representatives.

Performance Expectations

One could argue that the air transport system for reasons of increased security and economic needs also developed an increasingly complex set of performance expectations aimed at the consumer. The consumer must conform to a restrictive set of behaviour norms when moving in, through, and out of the aviation system. To start, passengers should have a common view and acceptance of airline staff authority. Next come exceptional skill, and an above average ability to adapt to automation. The preferred communication protocol with airline staff includes sequencing one's thoughts to facilitate the computer interface, and knowledge of the airline techno-lingo. In addition, passengers are expected to be:

- Knowledgeable about all passenger processing procedures;
- Able to orient themselves efficiently at airports;
- Able to communicate effectively with the staff;
- Cooperative;
- Compliant with all safety and airline rules;
- Attentive during safety briefings;
- Non-confrontational;
- Unaffected by fear of flying;
- Unaffected by nicotine and drug withdrawal;

- Unaffected by stress, fatigue, physical discomfort;
- Remain unaffected by a lack of information;
- Willing to communicate deferentially with the staff;
- Not needy for a quiet cabin environment, adequate space, seat comfort, food and drink;
- Remain buckled up, and remain seated throughout the flight;
- Be able to effectively evacuate after long-haul flights in crammed seating.

The primary need for such standard behaviour is linked to safety regulations established to protect the consumer from physical harm. Since the air transport system requires certain performance standards from its users, it thereby acknowledges that the passenger becomes, for a limited time, an integral part of the system. However assuming that passengers are able to readily comply with inherent behaviour expectations, is unrealistic. If passengers are to perform in a specific way when interacting with the system, it is incumbent on the system, to do all possible to facilitate a smooth integration.

Automation

The ongoing financial pressures on airlines for profitability and efficiency prompt growing automation of airline and airport processes, including security measures. The effects of automation on employee performance and its impact on the customer/server relationship cannot be overestimated as illustrated in the following example.

Automated Ticketing and Boarding (ATB) technology was initially designed to streamline the passenger process from ticket issuance to boarding by using a single document for each flight leg with a magnetic strip on the back. The benefits of this technology are multi-fold. Introduced at select airports, the airline customer with only carry-on baggage, can now bypass the check-in lineups, and proceed directly to security to the boarding gate. The use of magnetic encoding acts as a portable flight information database for each passenger that can be read by and used by each airline representative involved with passenger processing. It facilitates passenger identification, and has resulted in much-improved passenger and baggage reconciliation. Linking this system to airlines' revenue accounting is another major step to lower operating cost. Fraud prevention is an ongoing concern of the industry especially in light of electronic ticketing (ET).

Advancements in this area show industry's efforts to address increased demands for passenger convenience, security, and the need to reduce costs. Information technology is also a major factor in forging global alliances providing major advantages ranging from shared knowledge of its customer base for marketing purposes to shared knowledge of potential risk factors associated with passengers.

Reshaping Business and Relationships

Electronic access via the Internet to the airlines is a fast growing delivery channel for direct ticket sales to the consumer (Bowen & Headley, 2000). This is another cost saving strategy by the airlines to circumvent costs. This pleases a growing consumer sector, but results in de-personalized service with the potential for general alienation.

The developments in electronic ticketing play a major part in reshaping the way industry is doing business as well it is reshaping its relationship with its customers. It is moving from a customer/server relationship to 'client services'. This means in simple terms, instead of facing a live representative of the airline along key points of the passenger process chain, the air traveler will increasingly face self-service kiosks. Aimed at the computer-literate customer, they facilitate entitlements from electronic ticketing, and grant direct access to the departure lounge.

This development, although positive in many ways, is based on a flawed set of assumptions regarding the air traveler at large. Assumptions are made that the air traveler does not suffer from fear of flying, can read, is computer literate, is physically and mentally fit, is not stressed and has a degree of situation awareness[4] that guides him/her effortlessly through the airport. On the other hand, safety requirements and contract terms are no longer provided. This results in a new set of problems for information provisioning and passenger needs.[5] Since airline employee interaction with the customer will be virtually eliminated and replaced by new customized Passenger-Centric Equipment (PACE),[6] a term introduced by Industry, causes for frustration are shifted but not reduced.

Electronic displays are competing for passengers' attention without due regard for their ability to easily scan and select what is of importance to them. The lack of internationally agreed upon standards for airport signage does not address passenger needs.

A Question of Responsibility

The questions are how are casual users of the airlines going to adapt to this type of highly technical environment, and whose responsibility is it to facilitate the transition? The danger of alienation, mounting stress,[7] a feeling of disorientation and incompetence is real for the global air traveler with little education and experience in such an environment. The elderly, people who are under severe personal stress, the mentally challenged, first time flyers, people from third world countries, people with language problems, people whose coping mechanisms are not equipped to deal with the complexities of such an environment are subjected to increased social and psychological stress.

It is clear that the industry is in a continuous and major transition to automate its processes and information delivery. As global consumers are progressively exposed to these changes, they will gradually adjust. We can anticipate consumer generation gaps and resulting difficulties in coping within the highly technological environment. As in any well-managed change process, the transition phase must be addressed with full appreciation of the impact of these changes. A large and growing consumer group needs face-to-face interaction with an airline representative to cope rather than being forced to seek answers from an automated system. It is incumbent on the airlines to assist their customers in finding a comfortable balance between interacting with industry's technology and its workers.

An Alien Environment

The air traveler, unfamiliar with the technology of the industry, falls victim to greater fear and anxiety, uncertainty and mental workload in an alien environment. The accumulation of stress affects the ability to integrate the information needed to proceed to one's flight with ease. This can be measured by the type of announcements airlines make to call their passengers to the departure gates and the number of times passenger miss flights although they are in the terminal.

There is information overload, a myriad of signage, and advertising. Add to this, layered sounds of passenger vehicles transferring passenger to and from gates, their warning beeps, announcements, video machines, people talking loudly on their cellular phones or with each other without realizing or caring about the intrusiveness of their voices. This fog of sounds makes it hard to focus on information that is specific to one's own travel needs, especially if one is stressed or hearing-impaired.

A Question of Cooperation

The industry debate about efficiency and cost reduction does not take into account the unique cooperation it increasingly requires from its customers to maintain and improve its safety record. What are its plans to help its customers from all levels of society and educational background, to become compliant if not temporary players in this system? What is the role of the next generation of airline employees in the airports? Are there going to be any? Who do you turn to when you need help with the Passenger-Centric Equipment? Who do you turn to for information concerning delays and missed connections? These are situations in which the air traveler wants the assurance of a real human being, to air their frustration and obtain satisfaction.

At some airports in Canada, volunteers have been engaged, predominantly seniors, many of them retired airline employees who greet arriving passengers and offer assistance to anyone as required. This service is not only benefiting the air traveler but also acts as a source for alerting airport enforcement to potential problems.

There is little evidence that those with power to change the system are addressing these questions. What is their vision concerning the value of a caring, knowledgeable airline employee attending to the psychological needs of the harassed air traveler? What does the stressed air traveler do with built up anger without the opportunity of early intervention by an airline professional? Are we going to see a greater number of incidents related to denied boarding due to intoxication? Will there be an increase in vandalism of airport equipment? Will the social atmosphere on board be more explosive and more difficult to manage because of these changes? Currently there is no single study dealing with these issues, nor do individual airlines or IATA on behalf of its members support any research that might lead to a better understanding.

Consumer Advocacy on the Rise

Especially in the U.S.A, consumer advocacy groups are showing discontent with the aviation system at large. Targeting specifically industry's quality of communication, over-booking policies, fare structure, operational decisions, increasing cabin crowding, employee attitudes and service, the number of complaints has never been higher. At a time when the rift between consumers and the airlines widens, airlines demand that the consumer follow strict behaviour codes when in their care or else.

Frustration with United Airlines resulted in a class action lawsuit[8] filed on behalf of 168 passengers who were stranded for six hours on their flight on Christmas Eve 1997 at Milwaukee Airport. Passengers claimed they went without food or functioning washrooms before their flight was finally cancelled.

The settlement provides each passenger US$500.00 in cash, a US$500.00 airline voucher, and reimbursement of claims up to US$200.00 for unverified losses incurred.

Consumer activism is strongest in the U.S., where concerned travelers have aired their grievances in congressional hearings throughout 1999 demanding a 'Passenger Bill of Rights' after experiencing excessive flight delays. Delays have risen nearly 50 percent during the last five years.[9] Aviation analysts blame the overburdened air traffic control systems that are incapable of keeping pace with the traffic increases of the past few years and will not be able to cope with the huge forecasted growth in air traffic over the next ten years.

In the U.S. alone the FAA[10] predicts the total number of domestic traffic will rise from 604.1 million passengers in 2000 to 927.4 million in 2012, an annual growth of 3.6 percent per year. International enplanements on U.S. carriers are projected to increase from 54.6 million in 2000 to 108.4 million in 2012, an annual growth of 5.9 percent each year. Given these projected increases, the Air Transport Association views delays as a direct result of problems with the current antiquated ATC system and predicts a 250 percent increase if these problems are not fixed. Concerns that if the FAA air traffic control system is not fixed, the resulting delays will virtually eliminate the dependability of airline schedules and the system will descend into gridlock.'

Similar problems are evidenced in Europe. In 1999, schedule delays averaged 30 minutes, affecting a half-million flights in the first nine months of the year. These kinds of conditions are stressful for everyone and not easily rectified. Airline employees are routinely put to the test to deal with irate passengers. Passengers, on the other hand, object to less than honest and timely communication by the airlines, rude and uncaring treatment by the airline employees, compensation policies that are reluctantly applied, and then only when the customer makes clear and persistent demands.

No doubt, these conditions contribute to miscommunication, frustration, and increased stress on all sides. The airline employee and the customer more often than not enter a dangerous cycle where emotions take over, professional skills fail, abusive language strikes at times escalating into physical violence. Bor (1999) cautions in reaching premature conclusions:

> Attributing the cause of air rage to factors relating to the physical environment of the aircraft cabin or stress associated with air travel seems compelling but defies the available evidence and does not stand the test of face validity. Since most of these conditions have remained unchanged for more than two decades, one would expect every passenger to react to stress and their environment in a similar way.

Quality service research however provides some clues for customer discontent prompting a written complaint. Complaints usually are not triggered by a singular event of perceived service failure, but by a succession of service failures. Preliminary findings dealing with a similar pattern when examining incidents of passenger misconduct will be dealt with later on.

Social Safety at Risk

As air traveler misconduct is reportedly on the rise and in some extreme cases has endangered the safe operation of the aircraft, a socially safe cabin environment takes on a new significance. Such were the concerns of a captain on a Canadian charter flight in 1998, electing to refuse carriage to a customer with whom he had a confrontation on a previous flight. The passenger subsequently filed a complaint with the Canadian Transportation Agency. The complaint was dismissed based on the Agency's investigation (see *Appendix* A). The balance between individual needs and limited access to resources to fill those needs is difficult to achieve at best. The longer the flight, the bigger the plane, and the higher the risk for some situation to occur that contributes to passenger misconduct.

The social, cultural, economic, and educational background of today's long haul air traveler is highly diversified, adding to potential friction between fellow travelers and cabin crew members. Bringing together large crowds of strangers in the confined aircraft cabin presents different risks compared to small groups accommodated in a less confined cabin such as first or business class.

The sense of being recognized as an individual gets lost when one is treated as part of the anonymous crowd. Feeling they cannot be identified, passengers disassociated from their normal environment might choose to behave more negatively than in a small group situation where relationships are more readily established through individual recognition by cabin crews delivering more personalized service.

Group Dynamics in the Aircraft Cabin

Each time the aircraft doors close, it opens the curtain for a high-risk experiment of group dynamics. The aircraft contains an instant global village or, more accurately, a global prison camp. Picture this, the chances are that cabin crews face a group of people of which between fifty and sixty percent share a fear of flying.[11] All of them experienced a degree of stress prior to getting on the aircraft. Some of them have a personal history of violence. Some of them are on medication for a variety of reasons, including anxiety and depression, medications that magnify the effects of alcohol. Some harbor strong discriminatory views against authority figures, women, gays, people from minority groups, people who drink, people who smoke, people with poor personal hygiene, and the list continues.

There likely is the village pervert present, the village priest, the political big shot, the rebellious youth, the senile elderly, the village slut, and the village pimp, a new age prophet, a drug user/pusher, families and single parents, the ambitious career person, to name a few.

People travel because of their job, to go on vacation, to attend a wedding or a funeral, to care for a loved one, to immigrate to a new country, to apply for a job, to escape. They all rub elbows with each other. They can sense their differences and the seeds of stress and hostility have been sown. The crowded physical conditions the economy traveler finds himself in are extremely stressful. Social biologists tell us that invasion of requisite personal space produces stress and anger. It is, for many, reminiscent of torture, and wears down the thin veneer of civility. The potential for tempers to flare is ever present.

Taking Social Peace for Granted

Regulatory authorities and airlines take social peace in passenger cabin for granted. The reality proves to be different as passengers and cabin crew members will tell. Frictions between passengers are frequent, especially on full flights. Mini-wars can be observed between passengers negotiating space and elbowroom. It can be a subtle war or deteriorate into verbal exchange and shoving of seat backs into the upright position to open altercation. In countries where tribal feuds are a part of life, flight delays and incidents have occurred because heads of feuding tribes and their entourage faced each other off on the same flight, playing out their power conflict in disputes about their seating. Similarly, 'distinctive' groups who see themselves as 'special' engage at times in disruptive behaviour in the belief they can impose their demands on the cabin crew and other passengers. Chapter 6 deals in greater detail with risks associated with groups.

Economy of Safety

Airline policies and procedures, economy seating configurations and facilities such as washrooms and galleys are more often than not clearly designed to maximize economic efficiency. This is particularly the case with low cost operators and charter airlines.[12] Arguments by these airlines are typically: 'Passengers get what they pay for, they can't have cheap fares and premium service at the same time". The underlying rational, regularly trotted out, is, there are good business reasons to cram people into seats and provide minimum service: 'The consumer has to adjust.' While the quality of service is lowered, the organization's expectations of continuous civilized consumer performance remain the same.

It is not a coincidence that the topic of 'air rage' emerged at a time when airline service in general experienced a major decline. Airline Quality Reports compiled since 1991 in the U.S.A. and based on data published by the U.S. Department of Transportation monthly in the Air Travel Consumer Report, identify major increases in complaints since 1997: in 1998 the volume of complaints rose by 26 percent over 1997, and by 130 percent in 1999. Mergers, labour disputes and an outdated aviation infrastructure have exasperated air travelers who experience the negative effects of these developments that degrade the performance of the aviation industry.

Industry Complaints

Asking airline employees dealing with the public about their main complaints regarding passenger behaviour, the responses come quickly: 'They should know better than expecting us to perform miracles. They should be trained (in safety related behaviour and compliance issues). Some groups are very demanding; they all want to be treated special. People should prepare better for travel, i.e. bring their own medication, supplies for infants and small children, accompany elderly and challenged individuals. At special services desks, comments include: People with special needs come here with 15 minutes to flight departure and expect us to drop assisting another person at that time.

They require more time to get through the airport and to the aircraft, yet they don't plan accordingly.' Other issues are people's expectations of using carts to get to the gates. These are intended for the less physically mobile, however able body passengers frequently demand a ride, especially at large airports.

Regulatory issues such as restrictions on carry-on baggage also cause conflict between the airline workers and the customer. A ruling by the Canadian Transportation Agency defending the airline, illustrates this issue (*Appendix B*). Carry-on baggage requirements allow airlines flexibility in how these regulations are applied; this causes confusion and friction between the ground staff and cabin crew members while the passenger is often caught in the middle. Independent research conducted in Canada (DAHLBERG & ASSOCIATES, 1995) showed that over 95 percent of the respondents agreed with carry-on baggage restrictions while over 96 percent felt airlines should not bend safety rules for preferred customers.

Seating disputes are not uncommon. Some instances involve air travelers who feel entitled to be seated in premium class, adding their share to on-board conflict when they invade this section without clearance. Dress, personal hygiene, unfamiliarity with the aviation system, communication difficulties, and cultural biases are all part of the long list of complaints.

The list of complaints does not end here. Passengers at times, having ordered a special meal, change their minds and insist on receiving a meal from the regular flight menu. This causes problems when there are no extra meals available for the cabin crew to draw from.

Meeting System Expectations

A growing alienation between industry and its customers, and government attention to complaints reveals that integration into the system is far from smooth. Feeling alienated and stressed (Augustin, 1999, Loh, 2000), passengers' willingness and ability to perform according to expectations is severely affected. Conformity to airlines' expectations of customer behaviour is an issue that is gaining greater importance, as the aviation system is no longer capable of dealing effectively with passenger load increases and the consequent stresses on passengers. The airline industry is the only business in the service sector that imposes such rigid performance expectations on its customers. The overriding message is that airline employees would like a 'standard' passenger.

From the consumer's point of view the aviation environment has become increasingly impoverished. Cabin crews discourage passengers from getting up in flight and stretching their legs under the easily justified concern for safety, mainly incidents of clear air turbulence. Again the physical relief, these activities provide to the passenger are major, including reported benefits of reducing the risk of thrombosis. The combination of prolonged physical discomfort, and forced confinement to aircraft seats points to possibly serious effects on judgment and impaired performance.[13]

Complex Interface

The interface between the air traveler and the aviation system has become increasingly complex. Greater automation has given airline employees new tools to assist with the increasingly complex processes from ticketing, to checking in, to connecting on other carriers.

The change is visible in the amount of eye contact the consumer has with the employee during any interaction. It is considerably less than the time the employee requires looking at the computer screen while processing the passenger. Should the airline employee encounter any technical problem with the equipment, facial expressions and at times verbal commentary does little to assure the passenger that things are going well. Frowning, impatiently tapping the keyboard, sighing, and if all fails finally calling for assistance are all clues that add to the air travelers' increasing distrust of the airline's competence. The customer, looking for personal assurance and responsiveness from the airline representative, experiences instead a growing conflict between an increasingly alienating impersonal automation centered system and basic human needs.

Requirements

Since the aviation system requires basic performance standards from its customers, it thereby tacitly acknowledges that the customer becomes, for a limited time, an integral part of the system. The primary need for the air traveler's standard behaviour is linked to safety regulations established to protect the consumer from harm.

Much research has been produced on Human Factors[14] in aviation. However so far the concepts of this research have not been applied in the same way to the air traveler. Human Factors concepts provide insight into passenger behaviour and should be addressed in the design of airports and economy aircraft cabins with a view to maximize conditions for a socially peaceful environment. If the air traveler is to perform in a specific way when using the aviation system, then the system has to acknowledge the air traveler as a temporary member of the system, and do all possible to facilitate his/her integration into the system.

Up to now, the aviation system has failed to apply Human Factors concepts to the air traveler. This raises some important issues: what makes the aviation system so special that the same concepts cannot be applied to the consumer? Surely the consumer's brain, body, eyes and ears functions in the same way as do the pilots', air traffic controllers', maintenance personnel's, and cabin crews'. Surely oxygen intake, food and water, alcohol, drugs, and medication equally affect them. Noise, vibration, temperature physical exercise or lack thereof, fatigue, and stress too are known to place significant demands on performance.[15]

Investigation into employee performance takes Human Factors into account. No such allowances are made for the air traveler's performance. Applying knowledge of Human Factors to the air traveler is one way of reconciling the conflict between the aviation system and its customers. This approach may be helpful in creating a better understanding of the complex issues involved in leading to passenger misconduct, while simultaneously clearing the way to a new vision of what airline travel should be about.

The key issue for future safety in aviation is the system's willingness to integrate the customer by establishing conditions for a cooperative relationship. Predictions of an overburdened aviation system offer little hope for improving the relationship with the air traveler soon.

Consumer Research

One of the leading and internationally acclaimed comparative measurements of the quality of airline industry performance is the Airline Quality Rating (AQR) developed in 1991 by Bowen and Headley, researchers at the University of Nebraska and Wichita State University. The aim of this research is to provide an objective and consistent method for monitoring and tracking the comparative quality of major U.S. domestic aviation operations on a monthly basis.

The AQR is a weighted average of 14 elements important to consumers when judging the quality of airline service. The method takes published and publicly available data by the Department of Transport, effectively producing interval scale properties based on a mathematical formula, permitting a comparison across airlines and across times (Bowen & Headley, 1991).

It is not a coincidence that the topic of 'Air Rage' emerged in the U.S.A. at a time when airline service in general experienced a major decline. Airline Quality Reports compiled since 1991 in the U.S.A. and based on data published by the U.S. Department of Transportation monthly in the Air Travel Consumer Report, identify major increases in complaints since 1997: in 1998 the volume of complaints rose by 26 percent over 1997, and by 130 percent in 1999. Mergers, labour disputes and an outdated aviation infrastructure have exasperated air travelers when being at the brunt of these developments affecting the performance of the aviation community. These facts may well contribute to the rise in verbal abuse, however there is no hard evidence that these circumstances contributed to the more serious cases of assault.

There is also a growing body of little publicized research that gives a different insight into consumer behaviour. The airlines' traditional position is that the consumer is neither interested in nor worthy of being involved in more meaningful communication on safety issues or reasons for operational deficiencies. In fact consumers in highly industrialized countries are sophisticated and are more than ready to be treated with more respect and as an equal partner in the aviation system. They want open and frequent communication that can help and motivate them to greater cooperation.

The Becker Study

A white paper by Becker (1992) studied the concern passengers have about airline safety and offers recommendations of how to deal with the findings. This study is based on the premise that safety considerations play an important part in customer perceptions and need to be addressed in future marketing with the aim to achieve a more productive and profitable industry. The study reveals that four different passenger groups agreed on three major points:

- The availability of airline safety related information is insufficient
- Safety factors play an important part in choosing an airline
- Aircraft type was not an issue.

A total of 49 percent of the respondents felt that there was not enough safety information available with the majority expecting the government to provide this information. Becker points out the growing need for more meaningful safety information and proceeds to the rank the importance for obtaining such information to 14 sources. The major sources in descending order are newspapers, TV reports, personal experience and newsmagazines, word of mouth, and friends. By comparison, consumers rarely used libraries, government agencies, travel agencies, business associations, travel magazines, or travel reports. Becker further reveals a number of key factors with a high impact on passengers' feeling 'safe'. The factors are, in descending order:

- Safety information provided by the government
- Flying into less congested airports
- Greater number of air traffic controllers
- Newer planes
- Increased security
- Credible and readily available 'safety scorecard'
- Airline promoting their safety record.

New Information Sources

Since 1991, when Becker initiated his research, the introduction and explosion of the Internet has introduced a completely novel way of sharing information with users worldwide. The proliferation of web sites related to the topic of aviation alone is staggering. Regulatory authorities, investigative authorities, airlines, passenger associations, pilot and cabin crew labour associations, aviation colleges, news groups, airline manufacturers, aviation clubs, museums are just some of the major groups flooding the Net with information. A quick check in late October 1999 identified a total of 2,768,182 related sites for the aviation buff to explore. Surely, this has an impact on consumer awareness concerning aviation issues, including whetting the appetite for greater transparency on issues concerned with aviation safety. The consumer tends to prefer the easy to read and sensationalized reports with ready-made interpretations of this data to some of the primary sources mentioned below.

Sites produced by the National Transportation Safety Board (NTSB) in the US and the Transportation Safety Board (TSB) in Canada, are a good source for gaining access to accident statistics and accident reports. Similarly web sites produced by regulatory authorities such as the FAA (Federal Aviation Administration of the United States), Transport Canada, the Civil Aviation Authorities of the United Kingdom, Australia to name the more comprehensive sites, are designed to educate the public. The question remains, are these sites satisfying the consumer's need for the type of safety information they are predominantly interested when considering air travel?

Air carriers are now competing to attract customers with their websites. Northwest Airlines received the most top ranking awards for its website: in January 2000 it was named the top airline site by ZDNet and by Forrester Research, and received the number one spot in consumer confidence from the Gomez Advisors. Northwest is a leader in wireless technologies. Its website can be accessed by handheld computers and browsers, allowing the techno-savvy air traveler to obtain account information, flight schedules, safety information, special services, simplified airport maps for hub cities and links to CyberSavers, where the airline's low weekend travel deals are recorded. These awards are a reflection of significant management efforts to overcome previous trends of having the worst complaint rate in 1998 according to the 1999 AQR report and overall receiving the second lowest performance score of the ten major airlines listed.

The Dahlberg Study

DAHLBERG & ASSOCIATES conducted an independent survey on passenger expectations and perceptions of airline safety management with users of Canadian air carriers in 1995. We adapted the survey instrument 'SERVQUAL' to the topic of quality safety management. Although this approach differed in methodology and instrumentation from the Becker study, similarities emerged in a number of areas. The five quality dimensions of Reliability, Assurance, Tangibles, Empathy and Responsiveness were defined along the lines of safety practices observable by the air traveler and measured. Passenger safety management appears to be an issue for all respondents, and at times a key factor in choosing or defecting from an airline. Respondents commented on the airlines doing very little in promoting their safety culture and safety attitudes. The findings reveal respondents' concerns over airlines' commitment to the quality of safety management with major opportunities for better and more meaningful communication on safety issues.[16] Furthermore, respondents felt very strongly about passenger safety being a core aspect of the overall airline product. Contrary to the findings in the Becker study, users of Canadian carriers felt that the primary source for this information was the airlines themselves rather than the government since: 'they are in the business. They have an obligation to tell us, the customers.'

Suggestions to improve upon airline communication on safety issues addressed a number of areas. Safety videos received more favourable comments when compared with live safety demonstrations: 'I think the safety briefings on video are pretty good; at least they are consistent. Sure would like to see some humour injected, just because safety is a serious issue does not mean it has to be boring. They are not very creative in presenting safety information. They should get marketing involved. The safety features cards are accessible, but they are so busy. I do not know if anyone could make sense out of them if they had to in a hurry.' Other recommendations targeted the in-flight magazines for greater use of spreading meaningful safety information, using airport facilities and departure gates to facilitate and promote safety, including the installation of aircraft cabin mock-ups for public education.

Air traveler expectations and perceptions of quality safety management are precise and send a different message to the airlines than what their marketing departments assume to be important to the customer. How sensitive air travelers are to airlines' attitudes of quality safety management can be measured by the 80 percent success rate in soliciting respondents with an average time commitment of 20 minutes per person.

The survey respondents exhibited a great degree of awareness in safety matters coupled with a strong need to have their say. Many expressed dissatisfaction with being excluded from any discussions on airline safety. As industry and regulatory forums do not invite public participation they welcomed the opportunity to address this topic. A further benefit was the ensuing lively discussions on other concerns. On average, every fourth respondent commented on the need for tighter regulations to ensure airline safety. This was not surprising given the massive restructuring of Canadian air carriers at that time combined with deregulation and reduced government infrastructure affecting all departments, including Transport Canada. Respondents felt vulnerable and appeared ready for increased regulatory intervention based on their perception that Canadian carriers paid insufficient attention to the effects of their apparent safety culture. The user of Canadian carriers expects the airlines to demonstrate their commitment to safety consistently while applying the rules fairly to everyone.

Survey participants saw themselves as educated consumers of a product that promises safe and comfortable transportation. They are no longer in awe of air travel. They want the airline staff to be competent, conscientious, caring, and expect to be treated as full partners in the experience of safe and reliable transportation service.

The Polak and the Krajc and Pausch Studies

The Dahlberg study prompted further research by J. Polak for her thesis on the topic of 'Safety as Determinant for the Selection of an Airline'. The department of Aviation Psychology at Lufthansa Airlines supported this project. Polak focuses this study on the German leisure air travelers and their attitudes towards key factors used in deciding what airline to choose. Polak's work as does a subsequent study by Krajc and Pausch (1998) cover fear of flying and related research extensively. Fear of flying, common to more than one half of the flying public, may be heightened or reduced by the timely dissemination of safety information. Their conclusions offer practical steps for airlines to consider in designing appropriate communications strategies with the aim to reassure the passenger of reliable flight safety.

Almost all respondents (97.5 percent) expressed interest in the topic of safety. A total of 93.8 percent agree that airline should apply safety rules to everyone in support of maximizing flight safety (Krajc and Pausch 1998). These results are almost identical to Dahlberg's finding of 98.5 percent of respondents expressing the same opinion.

Polak addressed the issue of timely safety information in greater detail than Dahlberg's study. Approximately 1/3 of the respondents received their information first from the travel bureau they were dealing with. The issue of timeliness was an important factor. The issue of transparency was raised by 19 percent of the respondents concerning specific examples such as excess baggage, and size of carry-on baggage. Perceptions of German leisure passengers are significantly better than those noted by the Canadian study where a total of 54.5 percent of the respondent felt that relevant safety information was not provided in a timely manner.

In the area of service versus safety being a key determinant for customer decision-making, the responses from the travel agents and the customers could not be more opposed. The comparisons of travel agents' opinions on customers' preferences versus the opinions of the customers are of special interest. Out of a total of 9 factors travel agents rated in-flight service as the number one factor determining choice of airline while the customers rated safety as the number one issue in their decision, with on-board service coming in last. The age and the appearance of the aircraft rated second in the eyes of the customers and more important than the conduct of the cabin crews taking second place. Explanation of safety procedures ranked third in the customers' opinion compared to being ranked seventh by the travel agents. These gaps are significant and support the thesis that airlines and the aviation system are not in touch with today's customer needs, especially in relation to importance of safety.

Krajc and Pausch (1998) also focus their innovative research on 'The Perceptions And Significance Of Flight Safety From The Customer's Point Of View In The Deregulated Market'. They report the following:

The most impressive finding was the higher significance of flight safety compared to price. An overall evaluation with factor analysis was conducted resulting in the identification of 5 factors:

1. Service aspects
2. Concrete safety aspects
3. Safety image
4. Experienced anxiety during a flight
5. Decision criteria for the purchase of an airline ticket

These results strengthen the great emphasis on service aspects in the customers' judgement of flight safety.
This was one of the major assumptions the authors had at the beginning of the research. ...The results have important implications for airlines marketing strategies and service strategies.

The importance of Krajc's and Pausch's work stems from the development of a new survey instrument for measuring customers' perceptions and the significance of flight safety in purchase decisions. As with the earlier research by Becker (1991) and Dahlberg (1995), the findings are similar and validate the hypothesis that customer perceptions of airline safety are an integral part of the service experience, and as such have unexplored marketing potential.

Summary

Although there is no shortage of consumer research, most rely on infrequently and subjective consumer surveys that do not offer a basis for timely performance data and comparison. IATA as the industry lobbying association conducts extensive consumer research. However the results of this data are not easily accessible, and then only at a significant cost. None of the traditional surveys deal with specialized topics such as airline safety and communication with the air traveler. It is regrettable that the more creative research efforts remain sporadic and have not become part of the main stream.

Industry still fails to respond more sensitively to today's air traveler needs in providing more timely and meaningful safety information. It simply is not an issue. The marketing departments of the airline industry appear to resist the findings of the research mentioned above.

Industry does not have a stake in this type of independent research although the evidence points to a major shift of consumer interest in the topic of safety.

Safety and its potential for product differentiation by the airlines are missed. The training of employees would no doubt benefit from using these findings with the aim to achieve greater cooperation from the air traveler in complying with rules and regulations. Instead, the rift between the aviation system and its end users widens resulting in an ever-increasing confrontational relationship between the service provider and the customer.

Journey of Stress

The human factors' problems of air travelers are prompted by the need to adjust habits and adjust to a specialized environment. The following illustrates this point.

At home:
- Getting ready to leave, pack luggage, make arrangements for family members or friends, pets, make arrangements at work
- Leaving behind the known

Getting to the airport:
- Going to the airport, by train, bus – line up for ticket purchase, by cab, by car – line up to take parking ticket
- Depending on what type of air traveler is involved, this could be either a familiar or unfamiliar aspect of their overall travel experience

At the airport:
- Mandatory line up: ticket counter, customs and immigration, security, boarding, connecting baggage belt, line up for ramp transportation by bus or PTV (twice)
- Voluntary line up: newspaper kiosk, food and beverage concession, Duty Free, airport washrooms,
- Depending on what type of air traveler is involved, this could be either a familiar or unfamiliar aspect of their overall travel experience

Airport Line-ups

The number of times required to line-up ranges from a minimum of three to nine times, not counting line-up situations as a result of individual choices. Line-ups are undesirable experiences for the majority of people Not all line-up situations cause the same type of reaction. Depending on the air travelers' time constraints, the level of stress is either augmented or remains at a manageable level. General expectations include that the airport staff is aware that time is of the essence to most customers and staff is expected to act in a courteous and efficient manner.

Security Checks

The most common complaints regarding security checks focus on the bottleneck at security points and the security personnel itself. The air traveler with a high self-esteem, better than average income and position is particularly sensitive to having to surrender himself to professionals of lower income, social status, and different ethnic background. Comments range from 'they treat you like cattle, they show no respect they strip you of your dignity, they are arrogant' to 'they abuse their power. Other frequent observations include 'they appear bored', 'they act arbitrarily, e.g. two passengers with the same size and number of carry-on luggage are treated differently, one is let through, and the other is rejected or argued with by the very same security person.'

Should the alarm go off at the first automated scanning, methods for secondary security checks e.g. manual searches or the use of a wand, affect air travelers differently. Manual searches used in a highly industrialized country such as the U.K. prompt very different reactions compared to being submitted to the ritual of a hand-held electronic scanner or a wand. The intrusiveness of a manual search immediately triggers both physical and emotional responses. Individual reactions can vary greatly from an amused exhibitionistic perspective 'I found it rather flattering' to verbal and physical retaliation, as was the case in an incident at Heathrow airport in September of 1999, when an international celebrity protesting a manual search resulted in an arrest by police.

Customs and Immigration

Similar to complaints regarding security personnel and the manner, in which searches are conducted, complaints regarding immigration staff focus on their perceived abuse of power. 'It's a no-man's land, people take advantage of their position of authority, they are arrogant, rude, they belittle people, they have no manners', are some of the most commonly expressed opinions. The highest praise goes to Customs' set ups such as in Germany where the arriving air traveler chooses between a physical set-up of different lines with colour coded exits. Green indicates 'Nothing to Declare', red indicates 'Goods to Declare'.

The requirement for completing customs' forms is always considered a nuisance by those affected, despite progress made in simplifying the process, e.g. in Canada, one card serves for a family of six.

For the international traveler, the final act of passing through customs and immigration is often a final indignity and an additional source of stress. It is often experienced as having to submit one's individuality to an authority figure for which air travelers feel no respect. Passengers' lack of respect for these people is, in part, a consequence of their attitudes toward and treatment of the air traveler. Security personnel are perceived as lacking common courtesy, acting arrogantly, and stripping people of their dignity. Cultural aspects play a much more significant role in triggering these adverse reactions when associated with roles of authority. There are issues of reversed authority, when social norms associated with higher income and privileges are under siege at security.

There are race issues combined with language issues, as illustrated by this comment: 'The majority of the security personnel is either East Indian or Pakistani. I resent being bossed around in my own country by someone who is not fluent in my language.' The resulting hostilities are usually kept under control; they are preconditions to heightened stress and weaken coping skills. Human Factors considerations must become an integral part in the design of automation, processes and services with the aim to engage the air travelers' cooperation in achieving a socially safe cabin environment and the safe operation of the aircraft.

Outlook

The current outlook for air travel to double over the next ten years gives much concern to aviation experts and consumer associations alike. The failure by governments to modernize national airspace systems, and navigation systems to cope with the rapidly expanding air traffic continue to test industry as well as the consumer. Air Traffic Control and airport capacity are fully stressed. Increasingly passenger misconduct is an expression of overcrowding and linked to real or imagined service failure.

The relationship between the airlines and the consumer is already less than amicable, especially in North America. The latest annual surveys conducted by J.D. Power and Associates for readers of *Frequent Flyer* magazine and released in the Fall of 1999, reveals that one half of the respondents reported at least one bad flight experience in the previous year. Compared to other service industries, this leaves American air carriers trailing behind hotels and rental car agencies.

The Boeing Commercial Airplane Group also studies air traveler satisfaction. As reported in the October 25 issue of *Aviation Week*, Klaus Brauer at Boeing believes that on shorter flights, schedule convenience and on-time performance are most important to air travelers. By contrast, on long haul flights, the quality of inflight service and amenities play a more significant role. His research points to two major factors determining passenger satisfaction: on-time arrival and whether the adjacent seat was open. If these two conditions are not fulfilled, air travelers are more prone to criticize other aspects of their flight experience.

Can the loss of satisfaction on those two accounts be correlated to passenger misconduct incidents? My own research has not produced any evidence to that effect. The more intriguing question is why some cabin crews can deal with major operational delays and other challenges with seat duplications, meal shortages, and equipment failures without these situations culminating in passenger complaints or misconduct. I examined highly effective cabin crew teams and their leaders over a fifteen-year period and found some common denominators.

Preventive Strategies

In this section we will introduce a quality service strategy in support of a non-adversarial relationship and its integration with the self-protection model for cabin crews inherent in their safety role. Detailed operational strategies, specifically aimed at cabin crews in creating and maintaining a socially safe environment on-board, are discussed later in this chapter.

Winning Characteristics

On an individual basis, the best performers have previous experience in jobs related to the service industry or in businesses where complex problem solving is part of the workday. They value their job for the benefits they derive from it, the challenge of an ever changing work environment, a flexible life style, access to practically free travel world wide, and the opportunity to pursue their own personal interest in their spare time.

The star performers exhibit a keen focus on the task at hand, supported by their integrity in keeping up-to-date on changes to procedures and policies. Their balanced approach to safety and service goals enables them to make the right decisions when put to the test.

Overall, their interpersonal style distinguishes itself from those of the poorer performance because of its balance. They exhibit flexibility under trying circumstances. Problems are an opportunity to excel and hone their skills. They are self-assured, take ownership for the problems they face and act thoughtfully. Their emotional maturity sets them far apart from the majority of their colleagues, although they enjoy peer respect despite personal differences.

In their dealing with the public, they act with genuine caring. They understand instinctively the value of first impressions and the halo effect of helping one passenger in need, while others are looking on. They approach their job with efficiency, utmost courtesy, and always find something worthwhile to do, even on long-haul night flights when most passengers are asleep. They are good communicators. They listen, ask questions and seek clarification. They provide information and are sensitive to people's need to understand what is going on. People appear to relax in their presence. They exude quiet competence; they have a genuine liking for people, a desire to be proactive, combined with an ability to fit into any team.

I had the pleasure of meeting and working with many of such extraordinary individuals. Only an isolated few reported incidents of serious passenger misconduct during their careers. Most expressed the feeling that 'they were lucky'.

The Winning Teams

In addition to the characteristics of individual high performers, a good cabin crew leader enjoys the challenges of leadership and the personal satisfaction that rewards them after a job well done. I have observed two major reasons for superior cabin crew team performance. The first is related to highly developed leadership skills by the In-charge cabin crew member combined with operational competence. The second is related to the excellent teamwork by cabin crew members carrying a weak or incompetent leader despite all odds. The second case is rare and is more likely to occur in teams working on wide body, large aircraft.

Good leaders are respected for their ability to mobilize a group of individuals with varying personal backgrounds, seniority, different route experience, skills and motivation into a cohesively functioning group. This is particularly noteworthy since cabin crews do not necessarily work together for an extended period of time. Depending on their monthly schedules, the team composition can change not only from one flight cycle to another, but even within the same workday.

They tend to thrive in adverse situations. Unfazed by complex problems occurring simultaneously on different levels, they are able to prioritize quickly, delegate effectively and use all available resources at their disposal.

Good leaders share the following habits, they:

- Frequently communicate with their cabin crew and passengers;
- Level with them and solicit their support;
- Improvise on service and create options where others fail;
- Let their team members know where they stand;
- Clarify their expectations and follow-up during or after the flight;
- Value the contribution of their teammates and express their appreciation.

Self-preservation as a Motivator

Although formal or informal briefings are part of the aviation operation, they are often poorly conducted and fail to produce the intended benefit. Crew briefings and their effectiveness in dealing with emergency situations have long been established. The same principles apply to briefings on passenger issues. The problem is, how do you avoid complacency in delivery, keep creative and goal oriented? The following offers yet another approach to keeping motivated.

Public contact employees confirm, they have an important stake in reducing stress reactions to conflict situations and enhancing performance. Clinical studies (Taylor and Clark, 1986) and research in applied psychology support the value of 'preparatory information' (Inzana, Driskell, Salas and Johnson, 1996). Hardly anyone will question the benefit of preparation. Successful people excel at planning and organizing. Inzana, Driskell, Salas and Johnson define three types of preparatory information useful to consider when planning a briefing:

- Sensory
- Procedural
- Instrumental.

Sensory information deals with the unpleasant sensations, commonly experienced in high-stress events: anxiety, pounding heart, muscle tension, freezing up. Stress may result in perceiving a change in the quality of the environment when suddenly verbal insults and abuse explodes noisily in the passenger cabin. What was previously a benign atmosphere, transforms suddenly to a fast-paced threatening one. In many cases, cabin crews' individual and collective reactions are severely challenged. The awareness of how team members are likely to feel in a moderate or severe passenger conflict situation is helpful to counter any additional ill effects. The aim is to maintain effective performance of the team at large.

Procedural information describes the changes that may occur as a result of heightened passenger conflict. For example, this may include the reaction of other passengers and forms of well-meaning third party intervention, and the increase of time pressure on regular service functions. Steps can be taken in a briefing to ensure that unless help is requested, other cabin crew members continue their regular duties in addition to relieving those directly involved in the conflict of theirs.

Instrumental information addresses what to do to counter the unpleasant consequences of stress. Many airlines have formal Critical Stress Incident Debriefing programs in place helping their workers to deal with the consequences of having lived through an aircraft emergency or a threatening passenger situation.

Critics from cabin crew member ranks may point out that their briefing time is already insufficient for what they have to cover now. This argument is valid to a degree. Unfortunately, uninspired and poorly prepared briefings are the norm while superior ones are rare. It takes thorough preparation to make briefings meaningful, interesting, and motivating.

A commitment to a zero tolerance must be translated into action. Prevention is the key. Accordingly, a well thought out briefing is an important component to achieve this goal. There is no need to pack everything at once into a briefing. Since what is suggested takes making a change, addressing topics spread out over successive briefings is one of the better ways to accomplish this. Choosing a real-work example from the growing pool of passenger misconduct reports within an airline and industry results in a more creative briefing and has positive effects on reducing anxiety associated with these events. Naturally, the same approach can be easily adopted for team meetings with customer service employees working at airports. Making it part of the briefing agenda is an essential part of an overall prevention strategy. Obtaining valuable information on the range of reactions to be expected in their teams, lead cabin crew thus engages team members to actively seeking new solutions. The shared experiences are very useful in identifying resources within the team.

On-board Relationship Model and Explanation

The On-board Relationship model illustrates a framework to implement safe on-board relationships.

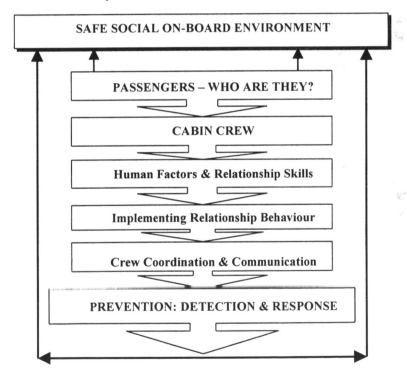

Figure 2.1 Human Factors Relationship Model

This model focuses on human factors influencing passenger and cabin crew performances. Cabin crew and air traveler demographics change on each flight and so does the social and cultural environment on board. Contributing to the social climate on board is the type of operation, such as a scheduled or a charter flight. The needs of the business traveler can easily clash with those of parents traveling with children or the vacationer. Sharing space at close proximity with total strangers can cause conflict, although most is related to harmless territorial jostling.

Six elements influence the management of social peace on-board. The first set of questions is, who are our passengers, and who are we?

Passengers are people with individual needs, unique expectations and little tolerance for service failures. Coming on board, they are already stressed. They paid good money for a service experience requiring their participation by following regulatory rules, airline rules and meeting variable cabin crew member expectations. Understanding these needs, cabin crew members are able to respond with compassion using on-board services as a tool and thereby accomplishing a safe social on-board environment most of the times.

The next step is to clearly identify what the main challenges are in establishing a cooperative relationship between the passengers themselves and the crew.

Knowing the Needs

Air travelers share very basic needs. They want to feel secure and they want to survive the flight. Mass travel ignites the intensity of these needs, partially because the physical environment triggers competitive behaviour. Physical space and comfort, privacy, stimulation, darkness, light, and silence are rare commodities. The Airline Pilot Association of South Africa (ALPA-SA) contends in a statement released in 2000 that long airport queues at airports and non-smoking policies have resulted in a major rise of passenger malcontent, often released on board the aircraft.

While traveling, body states too are off balance. People are hungry, thirsty, tired, too hot, too cold, experience a full bladder and colon, have tense muscles, feel ill, and are generally stressed.

After having accomplished the marathon obstacle course to the aircraft, air travelers expect to be able to relax and be taken care of. They prefer to avoid conflict and hostility. Mostly, they want to maintain their individuality, the prestige and status they enjoy in their professional and/or personal lives, and their comfort. Conforming to group standards and values of the service provider is difficult. It means abdicating one's power and accepting dominance from the service provider.

On a personal level they would like to avoid feelings of inferiority and failure in comparison with others or with their ideal self They want to maintain their identity while avoiding feelings of shame, guilt, fear, anxiety, sadness, anger, etc.

They want to be well informed of anything that affects their travel plans and physical well-being. They demand prompt, accurate and candid communication from airline representatives, especially when things go wrong.

Chapter 3 identifies in greater detail the physical benefits of communication. In the section on Consumer Research we gave examples of statistical evidence of airlines' failures to communicate effectively with the air traveler, especially during times of irregular operations, be they related to ATC, weather or maintenance.

Anticipating and Meeting Needs

Anticipating passenger needs is one of the most powerful strategies to create a socially safe environment on board. Airlines are not very good at expressing the value of service in a language that helps cabin crew members make the link to safety. Service is a strategy for self-protection; it effectively lowers the risk for social unrest. Cabin crews using their observation skills to pick up clues and act upon them before a request is made, are generally the best performing teams.

Some cabin crew members argue for example that: 'if I give one passenger a pillow (or blanket), they all want one. Our company does not provision enough for everyone, so we don't hand them out.' Although 100 percent provisioning is desirable from the cabin crews' point of view, it is not realistic. Management decisions need to consider equipment balance, competing demands on stowage bins' realty space, and economic objectives.

Discreet and selective tending to a person's need requires poise and confidence. Denying service to someone for fear of not being able to cope with subsequent requests shows little confidence.

Meeting the needs of others may mean a little 'extra effort' but it is also more satisfying and can be fun, while not meeting the needs triggers dissatisfaction and may lead to other problems. When denied a reasonable service, the customer's response depends on a host of other risk factors. Individual risk factors are covered in greater detail in Chapter 6.

Understanding Hostility

Aiming for social peace in the aircraft passenger cabin is important for reasons of safety, economics and competitive advantage. The astute airline will take on the challenge with a broad-based strategy and specific goals. One aspect is linked to educating airline workers on how human factors impact on passenger performance on-board. Passengers' rudeness and non-compliance to safety rules is not necessarily a deliberate affront to cabin crew members' personal dignity and authority.

Hostility is a symptom of a host of emotions. Fear of flying, the threat of losing control, fatigue, and personal and environmental stress are common experiences associated with air travel.

Alcohol and drug abuse or withdrawal lowers inhibitions and can lead to real rage. On-board smoking prohibitions may produce similar effects. Self-protection including from fear of flying, may be expressed in a number of ways: 'Why can't I drink as much as I want, have the seat I want, move my seat back, sit with my friend, smoke?'

- Defensiveness covers embarrassment. Cabin crew members' 'reading the rules to customers' risks escalation of hostility to aggression.
- A sense of helplessness relates to one's inability to change conditions or one's environment.
- A resentment of authority may be triggered by the image of someone with more power, influence or information to the point of hostility and anger.
- A person experiencing isolation is someone who is the only one who feels differently from the other people. Sometimes, a feeling of extreme isolation may cause overreaction and anger.

Considering the context of human factors may elicit more appropriate responses from airline workers. In the interest of on-board safety, service design and delivery can be powerful strategies to safeguard passenger performance.

The Cabin Crew

Cabin crew members' attitudes towards their job are crucial in establishing positive on-board relationships. Being aware of the duality of their safety and service role, and their personal communication style is a first step. Cabin crews are constantly performing in the eyes of the public, actors that need to obtain the public's respect, trust and cooperation.

Their role is not an easy one when facing potentially hundreds of different passengers during a given working day. Personal risk factors play a significant part in dealing with the demands of the public. Organizational culture[17] and aircraft environmental factors also impact on cabin crew performance as reviewed in greater detail in Chapter 6. Repetitive service demands, time pressure, and job routine, frequently erode cabin crew motivation, resulting in rote performance and complacency.

Supported by the International Transport Workers' Federation (ITF), Cabin crew labour associations emphasize cabin crews' safety role versus the service role. Unfortunately, the message is counter-productive. Air travelers are unable to appreciate the safety competency or the authority of cabin crew members unless they experience its benefits in an emergency situation. They are used to cabin safety routines performed with varying degree of commitment and style. Routine tasks do little to reinforce the importance of the safety role in the eyes of the customer. Demanding respect for a mostly covert aspect of their safety role reveals:

- Unrealistic expectations, since customers will not be able to judge competency in the absence of performance revealing special skills;
- The low value cabin crew labour associations place on the work of service providers in general;
- The depth of the role conflict and ambiguity they experience and so fervently seek to overcome;
- The lack of creativity in defining the role of the cabin crew professional of the 21st century.

The official positions are rigid while on the other hand many cabin crew professionals do not experience this conflict. They delight people under their care with genuine love for the work, their competency and charm, fluidly moving in and out of the various facets of their complex role.

Human Factors and Relationship Skills

Cabin crew members' emotional maturity, knowledge of customers' common psychological states and expectations, and detailed planning of each flight play an important part in developing strategies to affect a safe social environment in the passenger cabin. Recognizing cabin crew members individual differences, the In-charge will have to balance the various personalities, interests, talents and experience to maximize a successful outcome of a flight.

Communication with the air traveler is key to setting the tone for a flight. It has to be personalized. The use of industry jargon only irritates and alienates customers. Rote phrases immediately offend customers and demean the airline professional. Typical rote communication occurs when cabin crew members say their welcome and good byes; offer amenities, meals and beverages.

Skills in conflict diffusion, problem solving and crisis intervention are part of the working environment. The skills need to be well developed to cope with today's job demands and the diversity of the global air traveler.

Implementing Relationship Skills

Relationship skills are essential when dealing with people. That includes dealing with other members of the cabin crew. Air travelers are keen observers and feel assured by cabin crews who anticipate their needs and who act like a team.

Exceptional cabin crews understand service as a strategy to calm the nerves of stressed air travelers. Once the goal for a safe and peaceful cabin environment has been established during the briefing prior to a flight, all resources available to cabin crew members must be mobilized to that end.

Opponents of such an approach will identify short haul flights, multiple turn-around flights, time pressure and task overload as reasons to question the viability of such an approach. Based on my observation of high performing cabin crews, it is precisely their ability to create positive short- term relationships with new groups each time that is at the center of their effectiveness.

Crew Coordination and Communication

Crew coordination and communication, addressed in greater detail in Chapter 4, are key factors in achieving and/or maintaining social peace in the passenger cabin. The focus here is on planning the service as an extension of the safety briefing, assigning specific tasks to cabin crew members and 'passenger resource management'. The cabin crew briefing is the primary and formal opportunity to share relevant information from the passenger manifest or other sources. Special passenger needs and how they will be addressed should be clarified at this time.

Certain service policy and product issues with a potentially negative impact on passenger response are important to address and may vary according to individual organizations. Alcoholic beverage service for example poses some challenges to some air carriers, although not to all. An analysis of passenger complaints and incident reports at one airline revealed a lack of crew coordination and communication contributed to the majority of alcohol abuse incidents on-board. An example of this is related to passenger manipulation of the cabin crew. The passenger who has already been advised by one cabin crew member that he/she will no longer be served alcohol, approaches a different member of the team and obtains a drink. Cabin crew members are likely to inadvertently undermine each other's decisions, especially when these situations have not been addressed in a briefing.

Clarification of inter-cabin crew communication is important to avoid the hazards of miscommunication.[18] There have already been some cases where miscommunication between members of the cabin crew exacerbated a situation as illustrated in Chapter 4.

As part of the communication and coordination efforts, situation awareness and assessment and interpretation of passenger behaviour are important topics. What is upsetting to an inexperienced individual can lead to an over-interpretation of the severity of a situation when compared to the reaction and assessment of a more experienced colleague. To assist in achieving accurate information on passenger behaviour during emotionally charged situations, it helps to categorize behaviour as presented in the 'Conflict Continuum Model'. A reference to a level 2 or 4 rather than a narrative description of the range of behaviour expressions at each stage, saves valuable time and enables the In-charge to consider alternatives for intervention.

Prevention

Prevention starts with the company's zero tolerance policy. The qualities that contribute to its effectiveness, such as clearly defined management responsibilities assessing risks and associated losses, responsible service policies, products and processes, training for all employees in direct contact with the public, a comprehensive response strategy, should facilitate monitoring and measuring occurrences of passenger misconduct.

A survey of 206 airlines (Bor, Russell, Parker, and Papadopoulos, 1999)[19] investigated four areas of interest: requirement to record or report incidents of passenger violence; formal crew training to manage such incidents; policy permitting pilots to leave the flight deck to help manage such incidents; and passenger information regarding company's zero tolerance policy.

The findings reaffirm the need for a standardized method of reporting although 88 percent of the respondents confirmed a company requirement to record or report incidents. The Taxonomy of Disruptive Passenger Behaviour (Loh, 2000) is a major first step in closing this gap.

Although more than half of the respondents confirmed training was provided, the impact of such training on reducing incidents has not been established.

In the absence of clear and uniform policies, the issue of pilot involvement is very much open for debate.

The majority of airlines do not specifically communicate their zero tolerance policies to passengers.

Overall, these findings measure some progress since 1995 when the issue of passenger misconduct was formally tabled at an IATA Inflight Service Management Board meeting. Since the core issue of a commonly accepted methodology for recording and analyzing incidents has yet to be seriously addressed, statistics remain unreliable.

For the cabin crew, prevention starts with each individual. A holistic understanding of their role where safety and service functions are experienced as mutually inclusive overcomes role ambiguities that may negatively affect the interaction with the customer. Personal attitudes, value systems, attention to attire, demeanor and skills are prerequisites to succeed in this profession.

The In-charge cabin crew must have a clear idea what the leadership role entails. Demonstrating their own problem solving and conflict resolution skills when faced with challenging situations creates trust and instills confidence. A balanced leadership style that encourages participation, continuous communication, and flexibility is very effective in smaller working groups. A working group of more than ten cabin crew members requires a different approach. Chapter 4 deals with this issue in greater detail.

Seeking out complementary skills during the briefing is a very positive way of enhancing team performance. It serves to identify resources within the cabin crew team and allows for developing contingency plans should an irregular passenger situation occur. Flexibility of assignments, the identification of peer backup, encouraging cabin crew members to deal with a problem based on clear guidelines nurtures the development of needed skills.

Detection of Needs

Although service policies are intended to meet passengers' needs, they often fail simply because they are poorly communicated. Service enhancements are usually rationalized for competitive reasons but fail to make the link to cabin crew needs for self-protection.

Early detection of customer behavior signaling a need must be met with cabin crew responsiveness under normal operational circumstances to diffuse customer anxiety and frustration. As an important activity in the prevention of passenger misconduct, highly effective cabin crews treat it as part of their working routine.

In-charge cabin crew members who as a matter of course focus on observations during the boarding phase as a crucial first step in diffusing passenger stress and tension, tend to encounter fewer difficulties on their flights. They achieve a balance between regulatory requirements and ensure that the most stressed appearing air traveler obtains help.

The best performing cabin crews target high priority air travelers during that time and assists in whatever manner appears appropriate. Other air travelers observing these actions tend to relax. Observing positive steps taken by the cabin crew invariably increases confidence in the overall level and quality of service they are about to experience.

Cabin crews functioning in this manner find that they are able to reduce passenger tensions at an early stage and subsequently enjoy a better, less stressful flight. A non-confrontational approach achieves the best results. As one of the exceptional cabin crew members puts it: 'I never say 'no'. I always try to come up with an alternative. The passengers have to feel that I am on their side, and that I am willing to find a solution.'

Detecting Potential Passenger Misconduct

Again, crew observations during the boarding phase are important for detecting early trouble signs. Passengers who shove and push their way to their seats send signals of hostility. Others who are very flushed and perspiring might be under significant stress or be unwell. Eye contact with people in their vicinity, the manner in which they take their seats, help themselves to magazines or newspapers designated for premium class are all signs alerting cabin crew members to the heterogeneity of the travelers with whom they are dealing.

The most common reaction by cabin crews is to either stay away from these individuals or, at worst, to engage in direct confrontation. Some of the ways cabin crews retaliate against uncooperative passenger behaviour, are by immediately quoting the rule book and flexing their authoritarian muscle. This response sends a clear power message and tends to harden positions during such exchange. Quick judgments based on negative impressions are counter-productive and a sure path to escalating tensions.

The cabin crew, like any professional in the service or health care sector, must exercise control over their own interpersonal styles and maintain a flexible approach. Ultimately, de-escalation and self-protection are the goals. A situation that is not diffused at the onset due to cabin crew members' lack of professional skills, experience and/or attitude, may well lead to endangering co-workers and passengers alike.

Conflict Continuum

Table 2.1 offers an overview of the continuum of conflict and the behaviour associated with each stage.

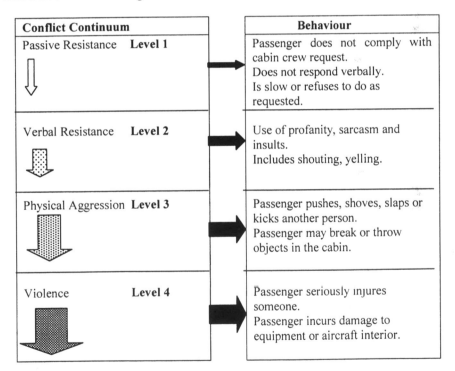

Conflict Continuum	Behaviour
Passive Resistance **Level 1**	Passenger does not comply with cabin crew request. Does not respond verbally. Is slow or refuses to do as requested.
Verbal Resistance **Level 2**	Use of profanity, sarcasm and insults. Includes shouting, yelling.
Physical Aggression **Level 3**	Passenger pushes, shoves, slaps or kicks another person. Passenger may break or throw objects in the cabin.
Violence **Level 4**	Passenger seriously injures someone. Passenger incurs damage to equipment or aircraft interior.

Table 2.1   Conflict Continuum

In a conflict situation, the behaviour does not necessarily follow the normal, progressive escalation pattern. A person who is predisposed to violence may immediately display verbal resistance, followed by physical aggression or violence.

The same type of 'leaping' can occur when there is a layering of stress-related pre-conditions that have exhausted a person's veneer of civility and is ready to lash out. This applies equally to the passenger and the cabin crew. Cabin crew members recognizing different personality types and differences in coping with stress are better equipped to respond to each. The primary goal is to stay in control and use tactics outlined later in this chapter, to remain calm.

Competency in handling difficult situations is a skill that can be learned and developed through practice. It is a life skill and a universal requirement for any individual working with the public, no matter what the industry sector is. The following model gives an overview of escalating conflict and typical behaviours associated with each stage. Level 2 to level 3 correspond in principle with the categories published in the FAA Advisory Circular (see *Appendix C*)[20] and adopted with minor modifications by the industry.

Critical Path to Disruptive Acts

Disruptive passenger acts occur as a result of three conditions interacting with each other: vulnerable passengers with predisposing traits, stressors and mitigating factors.

The following figure illustrates the critical path leading to disruptive acts:

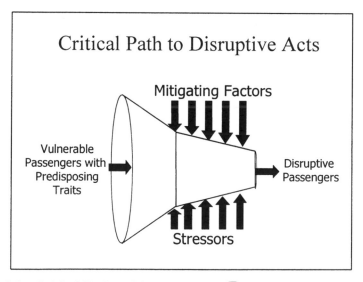

Figure 2.2   Critical Path to Disruptive Acts (Passengers)

Some passengers present more of a risk than others because of predisposing personality trends. These are covered in greater detail in Chapter 6. Vulnerabilities include e.g.:

- History of violence;
- Alcohol and drug abuse;
- Oppositionalism;
- Fear of flying;
- Fixation on entitlement;
- Low stress tolerance;
- Poor coping mechanism.

When these individual preconditions are aggravated by situational difficulties acting as stressors, the potential for disruptive acts increases. Stressors cover a wide range, such as:

- Problems at home, at work;
- Alcohol and tobacco dependency;
- Health problems, fatigue;
- Language difficulties;
- Purpose of trip;
- Airport processes and physical environment;
- Noise, temperature, vibration, fumes;
- Crowds and lineups;
- Flight delays;
- Lack of timely and accurate information;
- Uncaring and confrontational airline employees.

Fatigue is a well-researched issue with specific application to labour groups performing shift work.[21] Similarly, passengers having started their day tired or traveling long distances, often requiring the use of other transportation modes prior to boarding a flight, are prone to suffer from fatigue and exhaustion. Fatigue reduces the ability to concentrate, solve problems, and erodes the ability to be civil to others, especially in a conflict situation. Excessive fatigue also diminishes the ability to communicate and speak clearly, an aspect that may easily be misinterpreted by the airline representative.

Not every predisposed passenger is going to act on his/her frustration. Mitigating factors cover a wide range of conditions that may diffuse tensions created by stressors, and include:

- Being prepared: gathering information on airport facilities and processes, including airline check-in times;
- Awareness of self: how one functions physically, intellectually and emotionally under stress;
- Attitude: approach to safety procedures, a willingness to comply, to learn, to cooperate;
- Discipline: to act with reason;

- Being well-rested, well-timed consumption of nutritious meals;
- Knowledge: knowing one's rights under the tariff and airline compensation policies;
- Knowledge of legal consequences in case of disruptive acts.

Expecting passengers to perform as outlined above is clearly unrealistic although these are survival techniques that many travel savvy passengers have adopted. Learning from experience, benefiting from research and valuing the ability to cope without undue stress is the goal.

Given the benefit of well-designed airport layout and signage, accessibility to competent and caring airline employees ready to assist when problems occur, may mitigate stressors. Similarly, an aircraft environment with adequate seat configuration in Economy class, a less than full load, combined with a cabin crew attending to a passenger showing signs of distress, may diffuse tension build-up and avoid disruptive acts.

Dealing effectively with Human Factors for the sake of self-preservation is very important for the front line worker. Human Factors training is part of the education many airline professionals receive, especially crew members. Focusing on workers' performance alone ignores the relevance to passenger performance. If training were the solution, airline employees' attitudes and willingness to achieve a positive outcome in conflict situations would always be characterized by:

- Controlling own emotions;
- Staying calm, organized, and effective under stress;
- Being able to perform at a consistently high level for prolonged periods of time;
- Anticipating needs and respond in a timely manner;
- Recognizing different personality types and responds appropriately;
- Effectively dealing with the public in situations of distress.
- Diffusing conflict situations;
- Persisting in finding solutions or compromise acceptable to passenger;
- Responding to observations on symptoms of stress;
- Empathizing with passenger's plight, hearing what passenger means;
- Proactive communication;
- Cooperating with team members.

Human Factors alone cannot be expected to mitigating stressors associated with air travel. Airline policies and procedures relative to service with an emphasis on providing timely and nutritious meal and beverage service to offset any adverse health and mood effects are very important to ensure passengers' overall sense of well-being. Low blood sugar levels can cause weakness, irritability, difficulty in concentrating and, in extreme cases, unconsciousness. Adequate meal and beverage service should be considered as an overall preventive strategy; rather than overindulging one type of passenger and under-serving economy class passengers, a more responsible balance needs to be considered. Handling of irregular operations such as delays and compensation policies also either increase or diffuse stressors.

Response Techniques

The following techniques are helpful to maintain self-control when a person is engaged in a conflict situation:

- *Take a breath.* Taking a breath before responding verbally controls the voice and the blood pressure.
- *Acknowledge hostility.* Empathize, connect the person's anger to a point of view you can talk about.
- *Allow the anger.* Allow the person to be angry his/her way. Express no judgment in your verbal response or demeanor.
- *Don't take it personally.* Don't defend yourself against personal attacks. Get to a factual level. 'I know you want to ...., and not necessarily to have a personal grudge match with me...'
- *Empathize.* Find something in common: 'We both care very much about this situation. I see that you are very upset about this issue. So am I. May I offer you the following ...' Dilute the person's anger by offering a way out. By empathizing, you assure the other person that you are in control and reliable in your interpersonal skills.
- *Ask for clarification.* Asking for clarification focuses the other person on the issue itself. Taking the issue apart leads to a cooling-off process and results in suggestions for solutions and recovery.
- *Settle for disagreement.* Be willing, when all avenues are exhausted, to face the fact that the other person is unwilling to settle the disagreement. Demonstrate you logic, recap your position and the things you can do right now. The express compassion and understanding for their point of view.
- *Mediate conflict.* State the issue and tell why it has to be resolved. Ask Each party to state their views of the issue. Ask others, if necessary, to suggest ways to resolve conflict. Ask both parties which suggestions are acceptable to them. Other customers in the vicinity may also have to concur on the next step. Summarize agreement, and state how it affects your proposed next steps, if at all.

Cabin Crew Self-Protection Model

Table 2.2 discloses synergies between the traditional safety role of cabin crew members and the emerging new role involving a high degree of psychological skills leading to self-protection.

Current Scope: Safety	Future Scope: Managing Relationship	Training Points
A cabin crew ensures that air travelers are afforded an acceptable level of physical safety.	A cabin crew ensures that air travelers are afforded onboard social peace and emotional safety.	Cabin crew self-protection is the goal of safety regulations, safety equipment and procedures.
In an emergency situation, the role of a cabin crew is to provide leadership by performing certain duties and by directing passengers.	In a problem or conflict situation, the role of a cabin crew is to act with interpersonal expertise by using appropriate diffusion and recovery strategies.	Effective problem solving, decision making, and conflict resolution for self-protection.
To fulfill this role after a crash, a cabin crew must prepare to survive the crash in as much uninjured state as possible and remain in full control.	To fulfill this role during a conflict situation, a cabin crews must avoid hazardous attitudes and remain in full control.	Detecting risk behaviour. Estimate significance of risk. Identify options. Choose outcome objectives. Implement decision. Evaluate progress.
Pre-flight serviceability checks of safety and emergency equipment at assigned stations.	Passenger manifest and boarding phase are used to spot potential risk factors and identify plausible action.	Briefing to also focus on passengers issues, SPATS, communication strategies, and special procedures for passenger misconduct. Identify back-up.
Cabin crew is in full uniform.	Prepare to seek and give information.	
Cabin crew to be seated with restraint systems properly fastened before the take-off roll commences or before the aircraft is in deep final approach.	Task requirements and individual skills are matched.	Facing conflict: Know yourself. Take a deep breath. Identify hostility. Allow and understand anger. Get out of personal realm. Find something in common. Negotiate recovery.
Silent review.	Challenges: Their emotions, their defense mechanism, their perceived power – Your emotions.	Essence of conflict reduction: Diffusion of emotional energy, understanding differences and recovery.

Table 2.2   Cabin Crew Self Protection Model

Three elements should be given consideration when reviewing the evolving role of cabin crew members:

1. The Current Safety Scope entrenched in regulatory definition;
2. The Future Scope addressing work skills requirements in support of passenger relationship management;
3. Suggested Training Points.

Regardless of whether cabin crew members are experiencing real or imagined conflict between their safety and service role, success hinges on the proposed concept's benefit to them in dealing with their customers. The implementation of the previously discussed concepts should help cabin crews to perform with greater confidence and effectiveness, while at the same time becoming more creative and capable in dealing with difficult situations.

Achieving and maintaining social peace in the passenger cabin is in great part, the result of delivering service with a kind of artistry. Top management must be able to articulate how safety and service objectives blend together rather than being in conflict. To this end the design of policies, procedures, and services are important long- term strategies to deliver on the safety promise.

Cabin crew members are vulnerable to hazardous management decisions driven by predominant financial and marketing concerns. They have to deal with broken service promises first hand, while top management fails to accept their responsibility for poorly designed service systems. Clearly, they are no longer prepared to suffer the consequences of such a system. Just as industry decisions must be linked to growing air traveler dissatisfaction, so does intolerance by front line worker dealing with customer dissatisfaction.

Regardless of whether the corporate culture is sympathetic to human factors' issues relative to the air traveler or not, once the aircraft doors close, the cabin crew is on their own. The workplace does not provide any opportunity to remove individuals causing a disturbance. Where the cabin crew team includes a management representative as an additional resource, responsibility for achieving service goals still rests with the cabin crew.

Application for Customer Service Staff

The principles discussed in the On-board Relationship Model are equally adaptable to airline customer service employees in general. The following are the significant differences based on workplace characteristics and specific task context:

Physical Workplace

- Permanent and stationary;
- Affords some physical barriers between the customer and the airline worker;
- Electronic information systems are the main service tools;

- Should misconduct occur, help is accessible and escape routes are available;
- Allows for physical removal of passenger engaged in misconduct by others than the airline worker (e.g. enforcement, security).

Role

- Information processing principles drive role content;
- Labour contracts impact on task specialization;
- Efficiency standards and information exchange processes characterize the relationship;
- Safety role is minimal with no direct impact on own workplace;
- Social interaction is focused on one set of specialized tasks;
- Interactions with the customers are brief and transitory.

Because implementing relationship skills is condensed into a very short period of time, mental alertness, and articulate, easy to understand communications are critical skills when seeking and giving information. Service quality expresses itself in a combination of high tech and high touch. Assuring the customer with eye contact, a smile, acknowledging frustration and giving clear directions, is a positive strategy to reduce early tension the customer may experience.

People in line-ups observe how the customer service employees handle the air traveler in front of them and how effectively things are moving along. Positive or negative emotions may build and need to be acknowledged at the first point of contact.

No matter how short the interaction with the customer, detecting potential passenger misconduct at that time is critical. Should the customer exhibit hostile behaviour, signs of intoxication or other signs alarming the customer service employee, a communication alert should be triggered for assistance.

Assessing and determining if the situation is under control, are key to keeping incidents of passenger misconduct out of the aircraft and on the ground. Once the passenger confirms the conflict is resolved and is permitted to proceed to his/her flight, follow-up with the crew is a necessary preventive measure. Supervisor and managers may be called upon and in more serious cases the airport enforcement agencies respond to requests for assistance. The relationship between law enforcement at an airport contributes to the protection of airline ground staff; they keep potentially troublesome passengers off the aircraft. Examples in Chapter 6 illustrate just how effective a close working relationship can be.

It is human nature, especially under time pressure, to avoid getting involved in conflict situations. Traditionally, it is easier to hand the problem to the cabin crew: 'They have time to deal with this, I don't.' With growing awareness, crews expect more from their airport colleagues. Crews will quickly cast their rancour over their negligent colleagues. In case of dispute, the captain will make the appropriate operational decision. In the event a delay is called, management should reward decisions that keep passenger misconduct out of the aircraft, despite potential inconvenience, complaints, liability issues and other concerns.

Pentagon of Self-Protection

Just as athletes prime themselves for high performance physically and mentally, the public contact workers can benefit from a similar approach. The work is demanding, especially given the current state of customer malcontent. Performance lapses may contribute to escalating passenger aggression, clearly not in the interest of self-protection.

Figure 2.3 presents a set of critical questions public contact workers should address in preparation of their shift. The Pentagon of Self-Protection can be used for discussion in a team or used on a personal basis. The process is similar to that used in developing risk management programs at the corporate level.

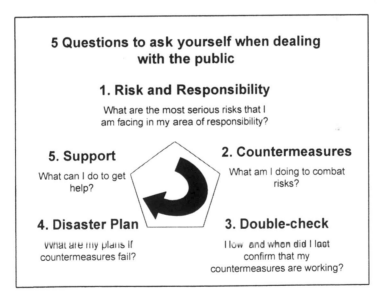

Figure 2.3   Pentagon of Self Protection

Explanation

1. 'Risk and Responsibility' focuses on risks relative to personality traits, attitudes, skills, training, type of risk situations, physical workplace characteristics, organizational culture, relationship with co-workers, etc.

2. 'Countermeasures' deals with decisions how to deal with the situation. These can be short, medium or long-term actions.
3. 'Double-check' involves soliciting immediate feedback from the person(s) involved in the situation. It also refers to activities within the organization aimed at identifying and reducing risks.
4. 'Disaster Plan' focuses on developing alternatives, using a range of behavioural responses and skills, identifying internal and external resources.
5. 'Support' deals with the identification and utilization of informal resources available at the time of an incident as well as the use of formal internal and external resources after an incident.

The previous examples are intended to offer models for training purposes. Motivation is the key. People have to feel that there is something for them, not necessarily for the customer. The Cabin Crew Self-Protection model suggests a new and expanded role for cabin crews, that goes beyond the regulated role definition. It focuses on integrating relationship skills and the requirement to spot symptoms potentially leading to passenger misconduct with the traditional safety role and aiming to diffuse such situations.

The airlines must ensure that their employees interacting with passengers are trained to treat passengers with respect and compassion at all times. Their skills need to be focused on prevention and detection of symptoms that could lead to escalating tensions. This is an area where workers interacting with passengers on the ground must play a much more active role. There are sufficient numbers of incidents involving serious passenger misconduct where there were clear warning signs prior to boarding and the problem was passed on to the cabin crew by the ground attendants, avoiding an unpleasant confrontation. Taking appropriate action at that time is critical in preventing incidents of escalated interference with crew members on-board.

Diffusion of tension through relationship skills is part of a self-protection strategy the same as standard operating procedures for emergency response assist crews to effectively deal with these events. To avoid being surprised and protect against ineffectiveness, airline employees interacting with passengers need a keen sense of situational awareness and maintain control. Training in these skill areas is essential to assist airline employees to cope with the daily occurrence of problem solving and conflict resolution.

Naturally, employees' attitudes vary greatly. External influences, for example airlines that are in merger situations, labour disputes or corporate culture that conveys an uncaring and contemptuous attitude towards their front line workers, impose significant stress on their employees. These stresses, especially when prolonged, can severely affect large numbers of employees in their ability to use their behaviour as a form of self-protection in the workplace.

For cabin crews, the uniqueness of the aircraft as a workplace in motion, detached from any organizational context and resources other than that of the crew team itself, self-reliance is essential. All manner of self-protection, especially in terms of professional behaviour under stress is a critical element in maintaining social peace on-board. Despite organizational or personal factors that may impact on a person's effectiveness, a focus on the primary motivation for choosing the most appropriate response is self-protection and self-preservation. This is a powerful link to maintaining control. Airline workers have to think differently about their customers. The framework of Human Factors allows for better understanding and supports social harmony on-board.

Airline Customer Service Commitment

Introduction

In the late 1980's a wave of quality service initiatives prompted by Jan Carlzon of SAS and known as the 'Moments of Truth' concept started major airlines to follow the trend. Economic recession, corporate losses and the need for competitive advantage were at the root of these initiatives. Corporate programs modeled after the notable example of British Airways, targeted front line employees. Some airlines hired media companies at great expense to motivate their employees with revival style meetings and one goal in mind: to achieve greater productivity and better service delivery to their customers.

These strategies have proven in most cases to be costly, of limited duration and value. Reward programs, although well meaning, created different kinds of problems with front line employees. Not surprisingly, in one particular case, resources allocated for rewards resulted in fierce arguments at the middle management level. Principles of fairness were given little consideration. Typically, the bigger the organizational unit, the more money per capita was allocated for employee rewards. Employees reacted quickly to the effects of unfair reward practices, presenting first line management with a new set of problems created by their executives.

Confusion resulted as to what was the essence of good to superior performance. Individuals known for their overall inferior performance would actively solicit commendations from customers, thus creating peer discontent and conflict for supervisors required to reward them based on the corporate program.

Airline management, especially in North America, has shown little ability to sustain a quality service strategy. According to consumers expressing deep-rooted discontent, lip service is being paid. Cost reduction and corporate greed are the guiding principles despite reported profits in recent years.

In the United States, the increasing complaints in air travel, escalated by the well-documented situation during a snowstorm at Detroit airport of January

1999, when hundreds of passengers were held captive on aircrafts for up to 8 1/2 hours, resulted in high-profile Congressional Hearings. Hearings in both the House and the Senate focused on the treatment of airline passengers and whether legislation was needed to protect air travelers from abuse by the airlines with the introduction of a 'passenger bill of rights'.

The discussions between Congress, the DOT, and the ATA led to the conclusion that, for the time being, legislation would not be necessary. Industry however needed to be seen as taking action. On June 17, 1999, Carol B. Hallett, President and Chief Executive Officer, Air Transport Association of America, signed a document prepared with its 14 member airlines, known as the 'Airline Customer Service Commitment. This Commitment would lead to each member airline to prepare a detailed Customer Service Plan by September 15, 1999, and full implementation by December 15, 1999.

Senator McCain, Chairman of the Senate Committee on Commerce, Science and Transportation, requested DOT's Office of Inspector General (OIG) to review and evaluate the progress of airlines meeting the provisions of the Commitment. Congress gave the DOT Inspector General the authority to conduct a review through The Wendell H. Ford Aviation Investment and Reform Act for the 21st Century (AIR-21), Public Law 106-181. Going beyond the original request, it expands the comparison of the customer service of ATA member airlines to include a comparison with non-ATA member operators.

Living up to the Commitment

The Office of Inspector General tabled its Interim Report on Airline Customer Service Commitment on June 27, 2000,[22] covering the first 6-month in their implementation phase. The report states that the commitment is in essence a recommitment to comply with existing laws with the exception of two provisions, one dealing with reservations, the other with an increase in the baggage liability limit. The Commitment and the Plans are lacking legal enforceability in contrast to the air carriers' contract of carriage, which, under Federal regulations is legally binding and as such are clearly enforceable.

The findings offer a clear overview of the discrepancies between the Customer Service Plans and the contracts of carriage, the DOT's capacity to enforce consumer protection rights, the importance of customer service in the marketplace, and suggestions for improvement. Furthermore, the airlines' practices concerning disclosure on overbooked flights and consumer access to comparative price and service information from independent sources are discussed while additional initiatives aimed at customer comfort and convenience are also noted. The report released on February 12, 2001 continues to be critical of airlines' failure to address causes of greatest customer dissatisfaction – flight delays and cancellations, including lack of adequate communication. Although this latest report notes some improvements, Congress's threat of 'Passenger Bill of Rights' legislation is still present.

The effects of Government cut backs since 1985 decimated the by more than half from 40 to 17 in 2000, while during the same period of time, the office workload expanded significantly. This is evidenced by the fact that air traffic more than doubled, and so did customer dissatisfaction as confirmed by an increase of complaints for the entire industry from 7,665 in 1997 to 20,495 in 1999. It is not surprising that under these circumstances, flight problems rank as the number one problem with 35 percent, and customer care as the number two problem with 20 percent. Customer care complaints include poor employee attitude, refusal to provide assistance, seating, and unsatisfactory food service. This alarming trend continues throughout 2001.

Timely and accurate information to the air traveler is a main issue (DAHLBERG & ASSOCIATES, 1995), especially in known delay situations, flight cancellations and flight diversions with concise reference to the airlines' policies for accommodating delayed passengers overnight. These issues are not new; these situations are not new; they are part of the operational reality of an industry affected by weather conditions, schedule irregularities due to mechanical problems or crew shortages, dependency on air traffic control and airport capacity, and last but not least on responsible management leadership.

The report reveals key characteristics of the mature aviation industry that go far beyond the periodic experience of the air traveler meeting an unfriendly airline worker:

- General reluctance to comply with existing regulatory requirements under the law;
- Lack of in-house tracking systems to ensure compliance;
- Taking advantage of DOT's inability to adequately monitor performance;
- Lack of accountability towards its customers and its front line workers;
- Disregard for the impact of its corporate values and decision criteria on social peace and its workers' ability to cope with increasing situations caused by service failures incurred, and regulatory requirements ignored;
- Shortsightedness in separating service from safety.

The level of commitment by the major airlines is questionable when existing obligations under the law are not met and result in consumer revolt. Crew members at the brunt of irresponsible management action, are endangered personally and make errors that threaten the safe operation of the aircraft in the process.

Adequate risk management principles appear to be absent. Furthermore, industry attitude towards regulators implies, unless forced by law, little voluntary action is taken to compete on the basis of customer service.

Employee groups interacting with the air traveler are treated with considerable disrespect when they receive inadequate training to deal with root causes of organizational neglect toward the customer. The message is: We don't abide by the rules, we make a mess, and you clean it up.

Overall these facts point to the conclusion that the already heavily regulated airline industry will not be prompted to change voluntarily. It most likely will have to accept further regulations that encroach on its defense to keep customer quality service issues out of the realm of legal liabilities. One of the consequences affects airline workers another affects air travelers. Both groups continue to be in a more vulnerable position since latent organizational and policy issues remain poorly managed.

A Link to Human Factors

The cost of neglecting existing regulations has affected the air traveler at the core of human factors issues: stress has been and continues to be placed to a high degree on the trusting consumer; inhuman conditions during long delays resulting in deprivation of food and drink combined with inadequate communication have taken a heavy toll.

The result is a consumer advocacy movement that is currently unparalleled, and is a clear signal to the industry and the regulators, demanding change. Advocacy is gaining momentum even in Canada, in large part driven by the effects of the Air Canada merger with Canadian Airlines. Complaints have tripled. Blending schedules, standard operating procedures, and workers from different organizational cultures is a formidable undertaking; any transition period involving such complexities will have its glitches.

The unfortunate by-product is the inevitable stress on the workers, not only do they have to face their own frustrations but also the verbal abuse passengers subject them to when things go wrong. The increase of passenger dissatisfaction is noticeable at Toronto's Lester B. Pearson's International Airport. The Peel Regional Police intervention during pre-board incidents rose substantially from the previous two years. Sergeant Bow comments: 'I'm fairly certain that this increase is an anomaly due mostly to the airline merger and associated problems.'

Without support and adequate resources from their organizations to ease the transition pains, employees are left to their own devices.

'RATER' – a Quality Service Strategy

The Five Dimensions of Service Quality

This section deals with a quality service model with the aim to demonstrate its effectiveness in analyzing complaints from a qualitative point based on the customer's emotional service experience. Although there is a wealth of service quality research and literature available, some are more helpful than others to move from theory to practice.

The most effective quality service model we have had the opportunity to work with was developed by Leonard L. Berry et al, renowned quality service researcher at Texas A&M University. Berry and his colleagues point out five broad dimensions that determine customers' experience of service quality. These five dimensions are particularly helpful in understanding what customers expect in regards to the emotional side of the service interaction and provide a practical framework for sensitizing airline workers interacting with the public. The acronym 'RATER' describes the five dimensions of reliability, assurance, tangibles, empathy, and responsiveness. Berry et al define these dimensions as follows:

- *Reliability* - the ability to provide what was promised, dependably and accurately
- *Assurance* – the knowledge and courtesy of employees and their ability to convey trust and confidence
- *Tangibles* – the physical facilities and equipment and the appearance of personnel
- *Empathy* – the degree of caring and individual attention provided to customers
- *Responsiveness* – the willingness to help customers and provide prompt service.

A Tool for Analysis

An analysis of complaints along the five dimensions of service quality gave us a surprising new profile of our cabin crews and their skill in relating appropriately to our customers. The questions we asked ourselves were, what is the first expectation cabin crew failed to fulfill in the customer's view, and were there any others?

In our case, cabin crews rated the poorest in the area of responsiveness. A total of 35 percent of the complaints indicated performance failure. Assurance and empathy emerged with the second poorest rating with 22 percent respectively. Service failures in the areas of reliability and tangibles were the least concern. The analysis further revealed that in 44 percent of the complaints multiple service failures occurred resulting in two or more quality dimensions lacking.

The findings of the In-charge cabin crew member classification showed a different profile compared to that of the rest of the cabin crew members. Empathy was the single most lacking factor, while assurance and reliability faired best. Not surprisingly, on short-haul flights, the lack of empathy in dealing with a customer problem was higher than on medium and long-haul flights.

Conducting a more detailed analysis of individual leadership styles in relation to the number and types of complaints led to another interesting discovery. Whether predominantly task or people oriented, if the style was strongly weighted, the number of complaints was also high. Since psychological testing was not used in the selection process prior to hiring these individuals, these findings surprised cabin crew members, and enabled them to develop strategies to achieve a more balanced approach.

A Tool for Change

These findings prompted the design of a new approach for focusing cabin crews on service improvements. The successful implementation required a carefully designed communication strategy and a commitment to involve cabin crew members along the way.

Because the In-charge cabin crews are so important in creating service quality on board their flights, they were the first to be briefed on our findings. As a result of consulting with them, they agreed to develop their briefings along the lines of 'RATER'. In doing so, cabin crews became more proactive and minimized their own work related stresses.

To overcome normal defensive attitudes was the biggest challenge. Previously, cabin crews received notification of a complaint in their mail folder with a request to respond in writing. Typically, responses indicated either no recollection of the incident, blaming another department or the passenger.

In recognizing the need for a non-judgmental and supportive climate to encourage professional skill development, we focused on determining diffusing techniques, preventing communication break down, and improving problem solving and conflict resolution. The aim was simply to coach cabin crew members to develop a broader repertoire of skills and experiment with different approaches.

Over a three-year period we issued monthly statistics on complaints and commendations combined with practical hints, often coming from cabin crew members themselves. This proved increasingly difficult in part due to conflicting priorities in an organization under severe financial pressure.

Labour negotiations with the cabin crew association in 1990 added significant stress. Despite the deteriorating relationship between the Union and management, our monthly statistics continued to be published to maintain a service quality focus. While other Bases experienced major performance deterioration during the labour unrest, at our location, cabin crews maintained a ratio of 7 commendations to 1complaint compared to 8 to 1 in 1989. No doubt, this was a magnificent achievement for our cabin crews.

A Measure of Change

In the first year, although complaint figures remained almost the same, commendations doubled, improving the ratio of four commendations to one complaint to eight to one. In-charge cabin crew members who previously had the highest number of complaints against their flights reduced these by one half. In the third year, a shift in complaints, eliminating empathy as a factor strengthened this achievement. A break through came with customers rating cabin crews' responsiveness and reliability as the leading qualities in their commendations.

Integrating the 'RATER' concept into all aspects of daily operation rewarded us with superior cabin crew performance and personal satisfaction. Moreover, coaching cabin crews honed our skills as well. We encouraged them to take risks and make decisions in the interest of the customer. For some it was a culture shock. At that time recovery policies were introduced, helping cabin crew members to make the transition. Some of these policies were confusing however learning from failures and celebrating successes led to achievements.

Some were able to perform miracles under the most adverse circumstances when everything went wrong; and still, customers sent commendations, rather than complaints. The 'RATER' concept focused cabin crew members to quickly identify customers' underlying expectations. As a result, more effective listening led to more appropriate action.

A Synergistic Safety-Service Model

Safety is defined as the condition of being safe, the freedom of danger or risks. Service as defined by L.L. Berry as 'the blending of server and customer performances, equipment, materials, and facilities.' The cabin crew's performance should aim at radiating assurances to their customers, persuading them through their demeanour and actions that they will be secure and safe while in their care. Ensuring compliance with safety rules must be achieved first and foremost in this manner; fuelling antagonism through intimidation is not a basis for a viable on-board safety culture.

Because marketing interpretation of service developed a more specialized view over time, it sometimes achieves nothing but a fatal cramping of service. Evidence to that effect is provided in Chapter 4. The basis for obtaining respect from the public is thus concentrated on cabin crews performing their service functions reliably, with empathy and responsiveness to individual needs.

Rarely demonstrating their full capacity as safety professionals, cabin crews need to be clear what aspects of their performance will gain public respect and trust. Research presented in this chapter supports our thesis that customers view safety as an integral part of the overall service provided by the airlines. 'Passengers feeling safe,' is in some part a function of the service culture in the cabin.

Acknowledging the generally high stress level, and the anxiety air travelers experience prior to boarding a flight, the first priority of cabin crews is to alleviate the stress as described earlier in this chapter.

The following model illustrates how the service role enhances the primary safety role of cabin crews in their relationship with passengers. The model is divided into two sections: the upper half describes the passengers' experience of cabin crews' overt service role, while the lower half depicts cabin crews' covert safety role. Passengers experience only a small percentage of cabin crews' safety role during regular operations while the technical expertise in emergency response and rescue remains covert.

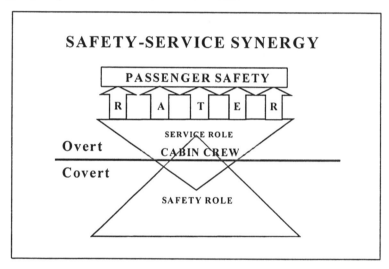

Figure 2.4   Synergistic Safety-Service Model

Safety-Service Gap

A safety-service gap occurs when cabin crews use their safety role as a primary means of interacting with passengers. This includes uncalled for authoritarian behaviour when dealing with minor cabin safety infractions during normal operations.

Requests for small service favours that can be easily accommodated are also frequently at the root of resulting conflict. Typical controversy swirls around the use of washrooms immediately after the seat belt signs are turned on and no turbulence is expected. Passengers' personal emergencies are occasionally met with rigid rule-bound responses. The passenger could have alleviated his/her condition and make it back to the seat in good time rather than being subjected to minutes of unproductive arguing.[23]

Cabin crew members performing their safety functions in a rote manner rather than assessing conditions at the time and making a reasonable decision is fraught with controversy. Airline liability and worker safety are at the heart of the matter. On the other hand, interacting with passengers in a predominantly authoritarian role will achieve an adversarial relationship with the public rather than one of cooperation and respect. The basis for obtaining respect from the public is thus eroded rather than strengthened.

Figure 2.5 illustrates the gap created by the role conflict between safety and service. As in the previous model, the upper half illustrates the passengers' experience of cabin crews' overt service role, while the lower half depicts cabin crews' covert safety role.

The inverted triangle represents a predominant safety role concept as a basis of interaction with the passengers during normal operations, while the service role is minimized. The arrows pointing up indicate the gap created by this role conflict.

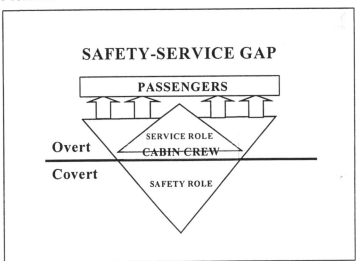

Figure 2.5   Safety-Service Gap

Challenges

The role of cabin crew members in the regulatory context is based on ensuring the physical safety of all passengers, especially in an emergency situation. Topics such as service obligations under the contract of carriage are not regulated for inclusion in cabin crew training programs. Compensation requirements and airline policies to this effect are usually deemed non-essential working knowledge for cabin crew members. The failure to educate cabin crews in these aspects leaves them at a disadvantage when dealing with irate passengers during long delays and irregular operations.

Prior to the regulatory requirement to carry cabin crew members on commercial flights, the role of cabin crew members was purely service orientated. Since then, the dual role of service provider and safety guardian has resulted in conflict between their safety and service role. Now, as a their role is undergoing another change, cabin crew members will be charged with greater responsibility for a relaxing and socially peaceful atmosphere on board. Customers do not wish to be intimidated by an 'airline warden'.

Any airline employee facing the public, be it on the ground or in the air, needs help in dealing effectively with the diverse customer needs. Passenger expectations of airline workers differ in an airport environment from that once on-board the aircraft. At the airport, proficiency in using the electronic tools of the trade serves mutual expectations of time efficiency. Worker-passenger expectations of what is to be accomplished are relatively homogeneous.

Passengers have a high stake in cooperating; the goal is 'to get through this is quickly as possible'. By comparison, passenger expectations on-board proliferate into a complex set of demands with a high emphasis on how the service is produced; it determines satisfaction or dissatisfaction. The need to cooperate and participate in achieving a satisfactory end result is no longer a prime consideration for the passenger at this time. As audiences of live performances, they sit back and judge the actors moving about in the cabins. They do not feel that they should have to deal with the consequences of 'backstage' problems such as:

- Crew shortages;
- Tired and stressed workers;
- Poorly trained workers;
- Provisioning shortages;
- Failures of the entertainment system, broken seat backs, plugged toilets;
- Dirty chair tables.

Engaged in repetitive tasks throughout their workday, workers tend to get dulled, even under normal circumstances. Complacency sets in, and one customer just starts to look and feel like another. To deliver service dependably and accurately each time is a major challenge.

Summary

Just as human factor issues at the airports and in the aircraft cabin must be linked to training, so must be a synergistic safety and service strategy. The quality of relationships between the airline service provider is fragile, and the increasingly adversarial trend alarming. Throughout this chapter I emphasized prevention. In a work environment where stress reigns, acknowledging it and using different ways of coping with its consequences is equally important for passengers and the airline worker.

Motivating oneself and others into performing with greater efficacy is key to inject some freshness into the old routines.

Traditional conflict resolution models are not necessarily effective in an aircraft environment. The environment in which the conflict occurs limits strategies aimed at diffusing emotional tension as compared to stationary and permanent workplaces.

Space restriction makes isolating the disruptive individual almost impossible. Single class, single aisle configured aircrafts are particularly vulnerable from that point of view. Conflict invariably is played out in front of an audience. This may result in overt or covert interference by passengers witnessing the incident. Task and time pressures curtail the cabin crew's ability to listen to the aggrieved customer's entire story, thus adding more fuel to the explosive emotions.

Employees are most likely to perform with greater confidence and efficiency when they experience the support of the organization in dealing with difficult customer service issues. Management has the responsibility to address root causes of role conflict and create a vision that integrates safety and service goals.

Airline executives can consider the benefits of their strategies by discussing the following questions:

1. What are the root-causes of passenger misconduct?
2. Do we have a tracking system in place for our customer commitments under the law?
3. How do our policies, procedures and service products affect the air travelers' ability to cope with our systems?
4. Do we accommodate the air travelers' need for reliability, assurance, empathy and responsiveness?
5. What are the most critical points in the process that cause the most serious stress for the air traveler?
6. What is the financial impact of passenger misconduct?
7. Is it a competitive disadvantage?
8. What can we do to reduce or eliminate those risks?
9. What would be the financial impact of reducing the risks?
10. What employee skill profile is essential to meet the social safety and security demands of mass travel?
11. What type of training would be required and at what cost?
12. How do we need to organize management to monitor and support our employees?
13. How do we measure our current performance and does it meet the changed requirements?

14. How do we implement and sustain a culture change that focuses on the passenger as a central human factors' issue?

Notes

1 A detailed account of one such experience is that of Renee and Michael Sheffer (1999).

2 The findings of a random survey (DAHLBERG & ASSOCIATES, 1995) demonstrate passengers critique airlines for poor management of safety information. The results show that safety information is important to air travelers, however it is generally poorly presented and timed. Moreover, the findings reveal a high degree of interest in safety matters, an aspect the airlines are not currently addressing.

3 Wood (1995) argues that aviation professionals are 'snobs', and deliberately exclude passengers from becoming active participants in the aviation safety system. He concludes that passangers are merely treated as 'cargo', and react by acts of minor non-compliance as evidenced in ignoring safety briefings. Outlining various methods to include passengers as members of the on-board safety team, Wood strikes at the core of the deep seated alienation between passengers and the aviation system. By acts of non-compliance, passengers seek acceptance from their fellow-travelers since cabin crews control every aspect of passenger activity on-board.

4 Situation awareness (SA) is formally defined as 'the perception of the elements in the environment within a volume of time and space, the comprehension of their meaning and the projection of their status in the near future' (Endsley, 1988). The air traveler requires situation awareness when going through an airport to the departure gate. At large airports, the multitude of shops, restaurants, commercial advertising and other facilities easily diverts passengers from perceiving the most relevant elements necessary to accomplish travel. Competing signage can result in confusion and a lack of comprehending where to go next. Visual images and noise fog blur comprehension and may render a passenger unable to decide what to do and where to go next.

5 See report (E.1.1.2b) by the Commission of the European Communities 'Protection of Air Passengers in the European Union' presented to the European Parliament and the Council in June 2000.

6 Terminology introduced by IATA.

7 Physical stressors are a challenge to situation awareness (Endsley, 1999), and so are social/psychological stressors such as fear, anxiety, uncertainty, mental load and time pressure (Hockey, 1986; Sharit & Salvendy, 1982). Although this research focuses on professionals in the aviation system, the findings can be extended to the air traveler who has to engage in certain performances prior to and while receiving the basic service he/she paid for.

8 *The Associated Press,* 7 October 2000.

9 *The New York Times,* 5 September 2000.

10 The FAA released these figures at its 26th Annual Commercial Aviation Forecast Conference on March 13, 2001.

11 Fear of flying in its varies forms is a condition suffered by a large number of air travelers with 20 to 25 percent experiencing significant aerophobia (Dean, Whitaker, 1980, 1982; Tortella-Feliu, Bornas, 1995; Metzen, 1991; Polak, 1997).

12 Beh and McLaughlin (1991) have researched aircraft cabin environmental factors on the rise of aggressive behaviour.

13 See Fifth Report on Air Travel and Health commissioned by the House of Lords and published November 15,2000.

14 ICAO circular #227 defines Human Factors as follows: Human Factors is about people: it is about people in their working and living environments, and it is about their relationship with

equipment, procedures and the environment. Just as importantly, ... its twin objectives can be seen as safety and efficiency.

[15] Factors affecting cognitive processing capacities change during the day as research has shown (Folkhard, 1990). Stressors such as noise, temperature, vibration, fatigue and individual differences affect a person's capacity for carrying out tasks (Bainbridge, 1999). Bainbridge categorizes five major groups for individual differences: Personality, interest and values, talent, experience, and no-work stressors.

[16] Arostegui, Pla, Rubio and Lorenzo (1997) point to airlines' Marketing and Advertising Departments' reluctance in referring to safety since it suggests 'the possibility of lack of safety'.

[17] Cabin crew teams are affected primarily by the culture of their association, their personal background and the influence of their seniority. They represent a subculture that can differ substantially from that of the organization at large (Dunn, 1995; Westrum, Adamski, 1999).

[18] Crew communication in case of emergencies is generally part of standard operating procedures (SOP), however is not standardized for communication involving service, passenger incidents and/or misconduct or medical emergencies.

[19] An overview of the survey findings were presented at the Eleventh International Symposium on Aviation Psychology on March 7, 2001.

[20] *Appendix C,* reprinted in full.

[21] Fatigue and biological rhythms affect human performance. Treatment of jet-lag, including the use of Melatonin and coping strategies (nutrition, naps, etc.) are the subject of a wide range of research (Akerstedt, 1990b; Arendt, 1988; Blake, 1971; Colquhoun, 1971, 1978, 1979; Comperatore, & Krueger, 1990;Costa, 1999; Dement, Seidel, Cohen, Bliwise, & Carskadon, 1986; Folkard, & Clark, 1993; Gander, Myhre, Graeber, Andersen, & Lauber, 1989; Graeber,1982; Haugli, Skogtad, Hellesoy, 1994; Klein, Wegmann, & Hunt, 1972; Knauth, & Rutenfranz, 1975;Koller, 1989; Lennernas, 1993; Romon-Roussraux, Lancry, Poulet, Frimat, Furon, 1987; Rogers, Spencer, Stone, & Nicholson, 1989; Sasaki, Kurosaki, Mori, & Endo, 1986; Suvanto, Partinen, Harma, & Ilmarinen, 1990).

[22] Report AV-2000-102, U.S. Department of Transportation, June 27, 2000.

[23] *Globe & Mail,* 19 August 2000: Actor Terrence Howard spent the night in jail after allegedly assaulting a member of the cabin crew during a dispute involving his attempt to take his 3 year old daughter to the washroom while the seat belt signs were turned on.

3   The Air Traveler – the Changing Face

Introduction

Air travelers mirror today's society, and so do their attitudes towards air travel; it is taken for granted. We are no longer in awe of the technology of flight. Authority figures have to earn our respect. We make no exception for crew members. Automatic deference is a thing of the past.

We scrutinize traditional institutions and discover their abuse of power. The pressures of disintegrating family structures combined with work pressures are measured by the increase in stress related illnesses. Bill Wilkerson, president of the Canadian Business and Economic Roundtable on Mental Health predicted a 50 percent increase in stress-related disability claims in the next decade. The economic impact alone is staggering with an estimated US$60 billion paid for cases related to depression alone. As workers face increasingly change in the workplace and pressures to meet often unrealistic deadlines, in Canada alone, the cost of mental disabilities is estimated at C$8 billion annually.[1]

As stress increases, so does workplace violence according to statistics. The Toronto-based Warren Sheppell Consultants Corp. is one of the country's largest employee assistance providers. During the period of 1995 to 1999 they have been called upon to respond to 15 murders and suicides at or related to the workplace. The International Labour Organization based in Geneva, conducts extensive research into workplace violence world-wide. Their findings show Canada, and not the United States, among the top five nations in terms of workplace assaults.[2] These findings however are not reflected in the tentative and incomplete figures reported in the Canadian aviation sector.

In this chapter we will address how airlines are managing the changed customer base. The question is, are current efforts effective?

A Snapshot of the Past

Airline workers who remember the 'good old days' lament the death of the 'civilized air traveler', a person of education, means, good manners, and suitable attire.

In the 60's air travel was still exclusive to the privileged and high-income earners. Caucasian men dominated the picture while women and children air travelers were rare. Airline restrictions for the infirm and people with disabilities curtailed their access to travel by air.

Conformity to social norms expressed itself in appearances. Suits and ties were the standard for men, while women's attire included hose, skirts and dresses, hats, gloves and handbags designed to contain no more than the essential female paraphernalia. Public life followed the social conventions of courtesy.

Conforming to airline rules was not the type of task it is today since there were only a few rules to contend with, using seatbelts for take-off and landing, putting up chair backs and chair tables. Overhead baggage compartments were nothing more than open racks. Aircraft accidents and investigations over the past thirty years changed all that. Regulatory changes as a result of formal recommendations tightened the safety net on one hand while simultaneously reducing air travelers' sense of personal control.

A Snapshot of the Present

Now, at major airports around the world, travelers reflect the cultural, ethnic and racial diversity we have become accustomed to seeing in many urban centers of Western societies. A myriad of languages mingle with the general sound vibrations at the airports. Airline travel as the leading means of global transportation tests our tolerance for individual expression at close proximity to the limits.

The topic of a decline in public civility appears of growing concern. In a Southam News poll conducted in Canada with 1,017 adults on December 27–30, 1999, 65 percent of the respondents felt that public manners would worsen in the next decade with a deterioration of road manners anticipated at 74 percent.[3]

A Question of Civility

Deteriorating civility is also evidenced in the type of passenger misconduct reported in the media. Amanda Holt, 37, and David Machin, 40 were fined a total of C$5,400 on April 5, 2000 in Manchester court for overt passionate behaviour on an American Airlines' flight from Dallas, Texas, to Manchester. The two passengers, both married and holding well-paying jobs at the time did not respond to cabin crew attempts to curtail their offensive behaviour.[4]

Although this type of behaviour is not new, cabin crew members appear to become more willing to deal with such misconduct that previously tended to be ignored.

The account of one business traveler who recalled an incident on a PanAm flight in the early 70s still felt very strongly about an event on a flight from

Frankfurt to New York. Apparently, a German sports team became all too familiar with the female cabin crew members who responded in kind. Resulting in abandoning of regular inflight service to majority of passengers, the In-charge did not take any action to put a stop to the inappropriate conduct. Although passengers lodged complaints with the airline, no follow-up was ever provided nor any compensation offered.

The airline manager confirmed investigating a passenger complaint reporting of being sexually harassed by a female cabin crew member traveling on a non-revenue ticket in business class. Similar cases of male cabin crew members harassing female passengers have also resulted in internal investigations and disciplinary action over time.

Some people feel their normal inhibitions do not apply once they are in a foreign social environment. This also includes sexual behaviour. Studies show for example that tourists of both sexes feel free to experiment while away from their normal social safety net. When certain tourist destinations and marketers use liberated mores as a specific focus of attraction, greater potential for passenger misconduct is to be expected, especially when liquor is a contributing factor.

National customs vary a great deal and the clashes with those of tourists can be significant. Some tourist organizations are becoming concerned about the negative impact of the culture clash. As an example, tourist boards of the cities of Barcelona and Bilbao have prepared leaflets on basic conduct to be given to tourists descending on their towns. Low-cost airline flights play a major role in the weekend visits to these destinations. Although there is interest in other countries within the European Union to see how effective this method is, opinions on this approach vary. Some hoteliers and airlines feel, this amounts to patronizing their customers and won't lecture them on how to behave at their holiday destination.[5]

Concerns regarding a loss of civility are not restricted to a particular sector of society. Workers in all service industries report an increase in customer rudeness, abuse and at times violent behaviour. Many feel that they are treated as second-class citizens.

The media uses term 'road rage',[6] originally coined by psychologist Arnold Nerenberg, to describe unsafe acts on the road and high ways. Even the term 'office rage' has been coined for uncivil behaviour between office workers and their bosses. Toronto writer David Kendall used 'urban rage' in an article on a lack of civil behaviour on city streets in the Globe and Mail in 1997. 'Rage' is used in connation with other inappropriate behaviour such as inconsiderate use of cell phones in public places. Putting media and its influence on using these terms to attract reader attention aside, criminologists, sociologists and health care professionals are very much aware that preventive measures are needed to help and educate people in better ways to dealing with stress and conflict. Conflict management is taking on a more prominent aspect in institutions such as schools, to help young people cope with a build-up of anger and avoid tragic outcomes.

Summary

Despite the reported deterioration of civility and the increase of stress and workplace violence, we also notice a growing trend in lower tolerance to such incidents. Organized efforts in preventing such events are spreading, providing support to business, workers, and individuals. Pressures on the public to stay within more acceptable norms of conduct are growing. The aviation system and its users are part of this trend.

There is now much greater recognition of the complex interplay of pre-cursors leading to passenger misconduct than just five years ago. Questions of corporate responsibility, values and economic considerations must be addressed in order to affect a greater degree of preventive measures and achieve social peace in the passenger cabin.

A Measure of Discontent

The U.S. Department of Transportation reported a 23 per cent rise in U.S. flight delays for the period of January to September 1999 contributed to a tripling of complaints compared to the same period the previous year. Bad weather, equipment problems and Air Traffic Control were the major causes for these delays.

Inadequate contingency plans and the lack of timely information trigger intense passenger emotions in major delay situations. This is a fertile environment to trigger confrontation, verbal abuse and in isolated incidents assault.

The seventh annual Business Travelers Lifestyle Survey conducted by the publisher of Frequent Flyer Magazine. Reed Elsevier Plc's OAG travel-information unit, and released in the Fall of 1999, reveals air travelers in the United States are becoming more frustrated with the airlines. The main reason for their malcontent is crammed seating, carry-on baggage management and record delays.

A new category dealing with 'air rage' was added in this poll of 3,000 business travelers in the U.S., Canada, the U.K. and 10 other nations. Some 38 percent of the respondents from the United States, or almost four out of ten frequent business travelers, averaging 20 trips a year confirmed they witnessed verbal or physical abuse against airline workers, almost twice the average in the survey.

According to statistics released by U.S. Department of Transportation in November of 1999, air travelers lodged 2,265 complaints against the 10 largest U.S. airlines in September, more than triple the year-earlier for the same month. On-line filing capability is partially contributing to the soaring number of complaints. The Federal Aviation Administration said in addition, bad weather and equipment problems are also at the root of a 23 percent rise in U.S. flight delays between January and September 1999.

The survey results further revealed that internationally, 40 percent of the frequent fliers witnessed at least one type of incident involving verbal or physical abuse toward airline workers and passengers, drunken conduct, unwelcome sexual advances or illegal smoking. More serious incidents of physical altercation represent approximately 1 percent. Respondents included frequent fliers from France, Spain, Germany, Italy, Japan, Hong Kong, Singapore, Australia, Brazil and Argentina.

In terms of awareness and publicity, the phenomenon of air rage has gained a steady momentum since 1996. Education of airline staff has been at the center of airline efforts, while the question of education of the traveling public is one still under debate. While in North America and in the United Kingdom regulatory authorities, enforcement and the airlines demonstrate closer collaboration in this regard, other countries are not as forthcoming. This is in part due to cultural differences and differences in the legal system.

Targeting Consumer Segments

Changes in society translate into every aspect of our lives, including the airline business. Airline Marketing departments are targeting the changing user profile with innovative products, specifically designed to acknowledge the diverse needs of their customers. This is evidenced by their choice of seats for premium class service, amenities for business clientele, the increased variety in meals and the efforts to offer a wide choice of in-flight entertainment. Trends in consumption of alcoholic and non-alcoholic beverages, duty free boutique products, reading material, and all other cabin amenities, are carefully monitored and assessed. These tangible products are the visible signs acknowledging the diverse customer preferences. They are costly items with an impact on the bottom line. Changes to the entertainment hardware also tend to affect the wiring in the aircraft and resulting maintenance cost. The perks used in the competitive world of aviation are not limited to the above examples.

Expressions of Conformity

A new kind of conformity has emerged visible in the new status symbols of our times. They include laptops, the latest model of cellular phones and personal audio systems. The choice of carry-on luggage, and footwear is a give away to spot the experienced traveler. The wearers and owners of brand name goods signal a caste system of their own. Groups have their own way of standing out of the crowd, appearance and demeanor easily catching the eye.

Expressions of Non-conformity

Today's travel attire shows little conformity to any dress code. Individuality reigns. Airports are a stage for observing the modern nomad. Wrapped in the cloth of their choice, people of all ages, and social means, people with visible and invisible disabilities, gather here. Some go for comfort, some go for national tradition, some go for show, and some prefer to display mostly skin. Attires do not signal a concern for the sensibilities of others nor do they reflect an understanding what might be more practical in case of an emergency landing.

Visible and Invisible Norms

Because visible and invisible norms of the aviation system are safety and process driven, air travelers who don't conform can expect to be ostracized and even punished. Carry-on baggage is one such example and one of the more contentious issues. Clearly, industry does not view carry-on baggage as the type of safety issue it should be. Regulators are sympathetic to commercial concerns as reflected in the varying regulations worldwide.

The final test takes place between the air traveler and the crewmember on board the aircraft. Crewmembers make their decision based on a number of considerations. These could include the flight's load factor, an assessment of how much stowage space is still available, preponderance for rules and consistency, including compliance to normative team decision making are some of the major ones.

Our own research conducted in 1995 revealed that air travelers using Canadian carriers are supportive to strict regulations. Respondents were however very sensitive to the inconsistent management of cabin safety regulations. In the example of carry-on baggage restrictions, a total of 40 percent of the respondents felt, employees were not enforcing the regulations, while another 39 percent were uncertain. In a subsequent item dealing with bending safety rules for preferred customers, only 11 percent of the respondent felt confident that airline employees were sticking to the rules. They expressed the opinion that if safety is the issue, regulations should be uniformly applied and should not be sacrificed for commercial reasons.

Institutionalized Processes

To travel by air means the forced immersion in a system of institutionalized transportation. There are rules to be observed for safety and security. Airline employees have unvoiced expectations of customers, their knowledge of the system, their courtesies toward them and other passengers. Conforming to rules is part of navigating smoothly through the system.

The rule driven environment creates a highly conditional relationship between the customer and the service provider and one that takes away one's feelings of control. Type A individuals are particularly sensitive to this perceived attack on their scope of control. This is not entirely the airlines' fault, as the unquestionable need for more sophisticated airport and security-oriented processes impacts on their overall operation.

Anti-consumer oriented rules such as disallowing carry-on food and beverages, limiting pre-boarding with children, restricting their seating to the back of the aircraft, constantly changing the rules of frequent flyer programs, are some examples frustrating the air traveler (Bowen & Headley, April 2000).

Deep discount fares attract many first time flyers that are less aware than average passengers of the rules governing air travel. Typically, conflict arises when Flight Attendants charged with the responsibility of enforcing rules are challenged or refused compliance.

Well meaning regulatory rules require Flight Attendants to refuse serving a customer alcohol beyond the level of legal impairment, and prohibit customers to consume their own liquor on board. Without the backing of responsible airline policies, procedures and training, these rules may in fact increase the risk of argument between the Flight Attendant and the customer.

Passenger attitudes towards authority and service personnel can be both hostile and demanding. Marlon Brando's famous line in the film *On the Waterfront*: 'I want what I want when I want it,' speaks for many of today's airline customers.

Access to Information

The air traveler like never before also has the opportunity to be better prepared before going out for a trip. Most major air carriers have web sites with specific information on passenger processes, services and helpful hints. So do regulatory authorities and other aviation advocates. Travel agents are another source of information should one not have access to the Internet. Although there is an obligation on the aviation professional to be responsive to the information needs of the air traveler, there is also an obligation on the air travelers to make use of this information in order to get a sense of control over the processes they must submit to. As the aviation industry is more and more relying on electronic distribution of information to its internal and external customers, air travelers without the means of accessing computers, are further disadvantaged.

Social Status – Risk Factors

Individuals enjoying the privileges of their social status can present a major challenge to airline staff working in a restrictive environment, governed by time and regulatory limitations.

Incidents where some politicians, diplomats, sports figures, entertainers, business, church leaders and preferred customers tend to insist on setting their own rules rather than becoming a role model and comply with regulations, attract a high degree of media attention. In a good number of these incidents, conflict may well have been heightened because of employee attitudes and lack of skills.

One of the examples comes from an American Airlines flight from Dallas to Miami on March 19, 2000. Tennis starlet Anna Kournikova, accompanied by her mother to participate in the Ericsson Open, reportedly refused flight forcing the pilot and police to intervene. The dispute occurred March 19 while Kournikova and her mother were on an American Airlines flight from Dallas to South Florida. The crew told police that Kournikova, 18 refused to put her miniature Doberman pinscher in its carrying case, as required by FAA rules. An argument ensued with the cabin crew, escalating to the point where the flight crew intervened to settle the dispute. Ultimately, the crew requested police to meet the flight upon arrival. The police report indicates that the passengers felt the cabin crew was rude and being obnoxious. No charges were laid.[7] [8]

Canadian Heritage Minister Sheila Copps found herself at the center of controversy on an Air Transat flight in early January 2000, involving a seat assignment problem, the airline had mishandled.[9] The media frenzy following the incident prompted the Minister to sue 16 news organization, Air Transat, Ogden Ground Services, and an unidentified Air Transat employee.[10] This case is particularly interesting since there are strong indications of miscommunication at the root of this mishap.

In summary, social norms of courtesy and a sense of social etiquette should be a standard part of cabin crew members' and airline workers' professional repertoire, even when rules are at issue.

Groups - Risk Factors

Groups, especially sports groups, transient workers, entertainers, charter groups, tour groups representing associations pose increased risks. One recent example involved the Pennsylvania Valley Dawgs after a basketball tournament loss on a Delta flight.[11] According to reports, a number of players refused to take their seats for take-off, uttered profanities, and argued when cabin crew members advised them that no further alcoholic beverage would be served. Group behaviour follows a different set of dynamics; group norms may be very strong and make it difficult to fit in the structured cabin environment. Group cohesiveness encourages offensive and rowdy behaviour.

One highly publicized incident dealt with the pop group Oasis who created a major disturbance on a flight on Cathay Pacific Airways with their arrogant and disruptive antics on a flight from Hong Kong to Perth in 1998. The group with their entourage of about 30 members apparently stood up on their seats, swore loudly, threw objects at other passengers and smoked on this non-smoking flight. The captain had to intervene in an effort to control this group and even threatened to divert the flight before some order was restored.

In another incident, a group of motorcycle racing enthusiasts returning on a Northwest Airlines flight from Memphis, Tenn., to Owensboro on October 2, 2000, created a fracas, throwing things at the cabin crew member, uttering obscenities, and intimidating the other passengers during the trip. According to reports, the group complained about the lack of drinks being served, the air conditioning, and the refusal to use the washroom.[12]

Similar incidents occur on charter flights, where air travelers are destined to vacation spots and get into a party mood devoid of any acceptable standards of conduct during their trip.

This is a global trend, having resulted in some airlines taking special measures to develop procedures how to handle groups with the aim to curtail misconduct on their flights and protect their employees. A statement by the Airline Association of South Africa (ALPA-SA) released in October 2000, addressing their concerns and that of passengers in that region. In particular, the statement focuses on sexual molestation, especially involving children. The association estimates that 50 percent of violent on-board incidents are related to smoking restriction at airports and in flight.

From Terminal to Transit Town

Airports should be built for speed, move them (passengers) in, move them through, move them out. The highly specialized and technical and artificial environment can be intimidating for the casual user. Airports around the world differ from one to the other, depending on the geographical location within a country, offering no additional conveniences to its customers, to highly developed major airports in the industrialized world.

Although much progress has been made in the redesign and modernization of airports, they still present difficulties for many travelers. Airports are confusing places, they are process oriented by their nature, despite much emphasis on customer service features. The newer ones resemble multi-purpose environments creating a transitory town with commodities ranging from shops to restaurants, bars, chapels, nurseries, therapeutic massage boutiques, beauty shops and hair salons, video arcades to the latest source of relaxation coming to Amsterdam's international airport, paid sex. Airports are converted into entities of attraction.

The diversification of commerce at major airports is prompted by the trend in airport privatization and the need to be profitable. The air traveler is a captive audience and waiting to depart is apt to spend money.

The bigger the airports are, the more intimidating they can be for the traveler. The urbanization of airports is not necessarily helpful in finding one's way to the departure gate. Unless the traveler is in good health and fit, is able to absorb instantly the myriad of signage does not require any special attention, and is not distracted by shops and all other commercial facilities, finding the right gate can become a major task. Add to this constant noise, frequent announcements, the danger of having one's space intruded by fellow passengers with body odours, constant pacing, talking loudly on their cell phones, harassed families, children running, laughing, crying, and the cocktail of stressors can be a potent mix.

Airports as Catalysts

Airports catalyze a whirlpool of human emotions in response to being immersed in staggering numbers of strangers in transit, while having to conform to the conditions imposed by the aviation system. The latest figures for 1999 released by The Airports Council International (ACI), the international association of the world's airports, reveal some staggering figures. Atlanta, Georgia, ranked as the busiest airport in the world with 77,939,536 air travelers. This is more than double the number of passengers having passed through New York and London, Heathrow, ranking number 20 and 21 respectively.

The ACI's prime purpose is to foster cooperation among its member airports and with other partners, including governmental, airline and aircraft manufacturing organizations. Through this cooperation, ACI makes a significant contribution to providing the traveling public with an air transport system that is based on the principles of safety, security, efficiency, and environmental compatibility.

Airports as hubs of masses in transit pose unique problems to airport management other than simply operating efficiently. Masses of disgruntled air travelers are one such problem and can lead to a crisis situation. Emotions are tested beyond the breaking point as observed occur during weather related delays that can bring a halt to air traffic for extended periods. Relationships are exposed. Tension, social pecking order, exhaustion, and excitement, sadness and grief, anticipation and joy, anger, hate, love and sexual prowess are all on display here. Behaviour is driven by one major force, me first. Civility is a fragile commodity. To leave as soon as possible is on everybody's mind.

The airport and its on-site workforce are also targets for criminal activities. Airports offer high-stake opportunities for crime such as theft and smuggling by organized groups involving airline employees, some of the primary concerns for airport enforcement in addition to keeping disruptive passengers in line.

Class Segregation

Airlines recognize the intrusiveness of a crowded environment and have gone to great expense to offer private physical spaces with a great selection of amenities, designed to afford the most valued customers comfort and serenity. The downside of this airline class system is the obvious effects on the majority of air travelers. With special check-in counters for the privileged for example, the people in the line-ups for economy class get a first taste of the diminished monetary value to the airlines. This reality can be especially unnerving when one observes an agent at the business or first class counters without any customers, making no attempt to lighten the burden of their colleagues. The more bold or desperate air traveler, seeking more timely attention will dare to go up to the better class agent. Observing the interaction, one often sees the agent listening, look at the ticket, and then simply gesture the rejected customer back to the economy line-up. Frustration is to be expected. The body language of both parties is expressive. The agent has established his/her special status and authority, in many cases well meaning and within the guidelines of the airline's policies. The customer, depending on temperament, may huffily or dejectedly retreat or in some cases, object vigorously to the treatment. All this takes place in public view.

Air Rage is Good for Business

The trend in catering to body and soul at the airports signals a new wave in entrepreneurship. At Calgary's International Airport, 'OraOxygen' is reported to be the world's first oxygen lounge at an airport. The owner, a cabin crew member with Air Canada, is familiar with oxygen deprivation on long-haul air travel and weary passengers. In addition to providing six breathing stations, complete with videos on proper breathing and relaxation techniques, air travelers can also refresh with a shower and enjoy a range of massages, manicures and pedicures. Expansions are already planned at three additional airports, two in the United States and one in Amsterdam.[13] With predictions by the Travel Industry Association of business travel to continue to rise, so should the demand of the wellness conscious men and women needing a quick way to re-energize before their next business meeting.

Service ideas do not stop here. A Dutch brothel chain offers relaxation with champagne, caviar, massages, and sex at Amsterdam's Schiphol Airport. What if any effect this new service will have on a decrease or increase in incidents of sexual harassment on flights departing and arriving at this airport would be interesting to explore at some time in the future.[14] Similar to La Vegas' airport, casino style gambling is available to capture air travelers' attention.[15]

Maine's Bangor International Airport, increasingly used for Atlantic flights to deplane disruptive passengers, has financially benefited from an average of eight to twelve flight diversions a year.

Airport Lounges

Airline airport lounges are special oasis for the privileged and crowd weary air traveler. These lounges are sanctuaries designed to please the most demanding customers and offer the latest in comfort and office technology. Depending on the airline, access to these exclusive spaces is based on the number of miles flown while others, less luxuries ones, are accessible to air travelers through bought membership. For the majority of air travelers, the discreetly marked doors never open but symbolize a harsh class division. Behind those doors there is peace and quiet, privacy and attentive service. Outside, the hassle and bustle of airport life with no private space to relax in before boarding a flight, engulfs the ordinary traveler.

British Airways has developed one of the most innovative approaches to combating the down side of air travel. The comprehensive program called the 'Well Being Service' extends from the aircraft to a unique arrival lounge at Heathrow Terminal Four. This lounge offers complimentary services aimed to sooth the travel weary customer. The environment features New Age aroma therapy and is designed to pamper the most coveted customers back to top conditions before releasing them to resume their hectic schedules. Spa users can ease the transition long haul frequent flyer customers have to make. It provides a special benefit in that it offers timely relaxation, immediately after arrival without the need to check into a hotel first.

Summary

Airports are places were highly technical systems blend with human expertise on one hand while on the other they clash with a myriad of expectations, emotions and needs of the air traveler.

Airports have a life of their own with its own rules and regulations. Public indignation concerning periodic security breaches is an indication of a culture of fear and discomfort associated with air travel. Airline professionals willing to accept this fact also know that countering these emotions with facts and figures will miss their target. An understanding and empathetic approach will go a long way to smoothing the transition for the air traveler.

Like never before, air travelers have access to a wealth of information, much of it electronically and designed to assist in preparing for their travel. Regrettably, there is no indication how much of this information is used in anticipation of a trip.

Innovative services at airports are quickly expanding with a view to catering to the stressed air traveler with airlines following this trend to maintain their competitive edge. It is difficult at this point to make any predictions concerning the value of these services long term.

The Changing Needs

Contrary to much mention in aviation circles of being a human centered system, customers are feeling more on the receiving end of an adverse and unfriendly system. Many customers are unhappy with the established aviation system. They feel its rules demeans them, that it imposes physical discomfort on the majority, diminishes personal control of their environment, and to top it of, makes them pay for this discomfort. Customers pay with more than money: they pay with stress, frustration, anger and, at times with medical and emotional implications associated with flying that extend well beyond the trip.

Is Bigger Better?

The airline business is costly with manufacturers seeking new ways to reduce cost for the industry. One such initiative has been under development for some time. Since 1991, Airbus Industry has been engaged in developing a new generation large aircraft, the A3XX. The concept phase terminated at the end of 1998, and more details are now being released to the public. The A3XX is considered a sub-group of the Airbus series with a basic version (A3XX-100) and a stretch version (A3XX-200). Combination freight and passenger versions, pure freighter aircraft are also developed along with short range, extended range capabilities.

The new generation of super-Jumbos will be capable of flying more people further than any other commercial airliner. At the 15th International Aircraft Cabin Safety Symposium in February 1998, Airbus representatives provided an overview of the safety related issues with a focus on design considerations relative to emergency evacuations.[16] The A3XX offers a seating capacity from 550 to 650 passengers:

The A3XX, is a four-engine, long-haul airliner seating 530-570 passengers in three classes - or up to 966 passengers in an all-economy layout. This is about 30 percent more passenger capacity compared to today's largest airliner, the Boeing 747-400.

The wide range of route capability ranging from short-haul high-density operations to very long trans-pacific flights, make the A3XX economically attractive. A turn-around time of 90 minutes - the industry standard for long-haul aircraft today - is considered essential with just two bridges joined to the main deck. It satisfies proposed ICAO Code F/FAA Group V airport requirements.

The two twin-aisled passenger decks are connected via stairs at the front and rear of the aircraft. The three class cabin layout features two aisles, with nine or ten abreast economy seating on the main deck, and seven or eight abreast economy seating on the upper deck.

With the increased air traffic, arguments for the utilization of a large aircraft such as the A3XX are persuasive. Airbus claims that direct operating costs per seat would be at least 15-20 percent better than that of today's largest aircraft. This is a fact the airlines consider very carefully when planning new aircraft purchase. Cost control while maintaining the continued fall in cost of travel in real terms are strong arguments in favour of the A3XX.

According to Airbus Industries' forecasts, there is a potential for more than 800 orders up to the year 2014 and 1,300 over the next 20 years. The entry date is now projected for 2005 to 2007. None of the North American carriers has officially declared their interest. Possible launch customers are British Airways, Air France, Lufthansa, Qantas, Singapore Airlines, Japan Airlines, Cathy Pacific, and Virgin Airlines.

The latest reports indicate a new level of comfort and luxury. There is talk of lounges, shops, a separate restaurant; children's play area and the latest technology, a shower that runs on five gallons of recycled water.[17] Surely this would not be the case for the high-density version planned to carry up to 966 passengers. Richard Branson of Virgin Airlines is reportedly enthusiastic, envisioning the new-generation aircraft as an airborne cruise ship with comparable comfort.[18]

Crew Communication and Crowd Control

Information on how crew communication will be structured on the A3XX is not available to us at the time of addressing this issue. As we discuss in Chapter 4 in more detail, cabin crew leadership and communication on incidents in the passenger cabins is of prime importance to either minimize or resolve events before they get out of control.

Scenarios of multiple passenger misconduct are more likely to occur on large aircraft. Questions of flight crew being apprised of such incidents in a timely and efficient manner are pre-requisites for maintaining social peace on-board. Already the demands on the In-Charge cabin crew members are significant on large aircraft, especially if they also perform a full service position on a bi-level aircraft with multiple cabins.

The prospect of the new-generation large aircraft such as the A3XX, poses new challenges to both inter-cabin crew communication and communication with the flight crew. A priority driven communication system may not adequately address the new reality. Even the concept of self-contained and managed sub-cabin crew teams is not fully satisfactory. Crew resource management will experience a new challenge for which there are no easy answers.

The new demands on cabin crew members' skills and performance necessitate a sober analysis, the development of a new profile and training focus for this workforce. Should airlines decide to add features such as exercise space, shop and restaurant, again new elements are introduced for re-thinking emergency procedures[19] and monitoring of an increasingly diverse aircraft environment. We anticipate a major change in terms of flight and cabin crew interface due to shifting social dynamics on-board with passengers moving about, freely utilizing different cabin spaces, resulting in much greater interaction with other passengers. Critics feel that these facilities are more appropriate for space stations; they argue, that airlines should stick to their core business as a convenient transport mode carrying people efficiently from point A to B. The visions of some have great potential to become a nightmare for flight and cabin crews working in this type of futuristic environment. Questions of handling sudden and severe turbulence and weight and balance issues come to mind when large numbers of passengers are not in their seats. Safety experts will no doubt be concerned about the vision of a 'Titanic' in the air.

Larger airplanes also mean larger number of passengers to be processed at the airports. Unless the aviation infrastructure at airports is upgraded, even the promise of enhanced passenger facilities onboard, are not reducing passenger malcontent. With the already overburdened aviation system in North America[20] and Europe, lengthy taxi-in and taxi-out times can be expected until governments and regulatory authorities take drastic steps in improving this situation.[21] As became apparent during the snowstorm in Detroit in early 1999, passengers can be trapped on an aircraft for hours on end. It is not difficult to anticipate what could occur when individuals start to express their extreme frustration and become leaders for the many others who share their emotions, resulting in passenger revolt on aircrafts capable of carrying an average of 600 people or more.

Aircraft Cabin Comfort

Compared to passenger comfort in the early days of civil aviation, today's high density aircraft configurations provide luxury accommodation. Sitting behind the pilot in an open cockpit, comfort consisted of leather coats provided by the airline, helmets, hot water bottles, and cotton balls for earplugs.[22]

Airplanes are primarily designed for safe and economic performance. Safety and economics are the driving factors for the enormous technical advances made over the last thirty years. Airline manufacturers have responded well to industry's needs and affected predominantly the operation of modern aircraft while in the process aiming to reduce human error in the cockpit. This internal focus has prompted an entire new wave of research aimed at the operating crew.

Technical aspects appear to be well covered as evacuation requirements drive cabin safety issues and research in the area of door and window exit design.[23] Installations of emergency exit floor lighting and cabin safety equipment are part of the response system in an emergency situation. Seat design must meet regulatory requirements, however seat density and low number of washrooms to high passenger ratio for example has not been considered a safety hazard.

High-density aircraft configurations with the aim to maximize profitability create conditions for heightened passenger tension. Cultural and individual differences regarding personal space definition and minimum tolerance will require a more prominent factor in aircraft configuration design. Airport capacity and air space management are critical for maintaining a satisfactory customer service level.

'Airlords'

Airline customers who pay full fare are the 'airlords' of the airline business; they make money for the airlines; their loyalty must be gained. They drive marketing to search out new and more alluring ways to please their needs. No money is spared and the competition for ideas from acclaimed designers is fierce.

The airlines determine what type of seat they are offering to both their premium and economy class passengers. There are no problems with space and comfort for the first- and business-class passengers, with airlines continuing to entice their most coveted customers with ever-greater luxury. The current premium seats are very versatile, the best featuring ergonomic design with full reclining capability, adjustable lumbar support, entertainment and office functions apart from the standard features, such as light control and call button.

British Airways announced on January 31, 2000, its new standard of luxury for the business traveler. One of its features is the Club World seat, by German racing car driver seat designer Recaro; Conran and Partners are behind the Concord's new interior, and Kelly Hoppen has given BA's first class a look "to emulate the timeless elegance of a Rolls Royce".[24] Amongst other amenities, the star feature is "the lounge in the sky", offering utmost physical relaxation and privacy to the business traveler, the most profitable segment of the market. Individual cabins, recreating the era of bygone travel splendour, contain a seat that folds out into a 1.8 meter flat bed was already introduced in 1996.

Long-haul travel in these types of seats caters to the air traveler's physical well-being while the economy passenger has to endure the same trip in an a mostly upright seating position with very little space to stretch their limbs.

Swissair competes for premium customers with its new cabin design by Swiss architect Tilla Theus. The new seats were inspired by a classic American design, the Eamus lounge chair to tap into people's need to be reminded of 'something familiar and comfortable'.[25]

Singapore Airlines has taken a different approach and offers 17.7-inch wide seats with a 32-inch pitch to its economy class passengers on its B747-400 aircraft. Additional features are individual video monitors, in-seat phones, and adjustable head and foot rests. In its latest move, premium seat on intercontinental flights will offer the state of the art technology with fibre-optic in-flight entertainment at a cost of US$18,500 a seat.[26]

American Airlines announced on February 3, 2000 a complete

transformation of its seating configuration in economy class. Responding to customer consistent demand for more space in this class of service, approximately 7,200 coach seats on its entire fleet will be removed at a cost of about $70 million. This move will offer increased seat pitch from the present industry standard of 31 and 32 inches to 34 and 35 inches, reaching as much as 36 inches on some of their aircraft. The first converted aircraft, a MD80 went into scheduled service on February 12, 2000. The seat conversion on the entire fleet of more than 700 jet aircraft will be accomplished within two years.

This trend of expanding space for economy class travelers is a calculated business decision and puts pressure on other carriers to follow suit in order to be competitive. With more space comes also a new economy seat that is ergonomically designed with six-way adjustable headrests, and power ports for computers in selected rows.

We see this development as a major breakthrough since the physical improvement will positively impact on the air traveler's feeling of comfort and reduce stress. It would be interesting to track the number of incidents relative to passenger misconduct on the converted aircraft for analysis and comparison with those occurring on the unconverted ones.

A more controversial issue is the use of cell phones on board aircrafts. Virgin Atlantic Airways announced 'Earth Calling' a service available on one of its B747 aircrafts allowing passengers to receive call on their mobile phones while in flight. This involves receiving call ringing quietly on passenger earphones or register in text on the in-seat video screens. To take the call, existing telephone handsets at the seat location are used.[27]

Creature comfort even extends to the cherished pets of airline customers. Northwest Airlines is introducing its innovative 'Priority Pet' program by adding 18 positions at three of the airline's hubs and a toll-free number for pet owner assistance. Pets will be transported in air-conditioned vans, and their owners can choose between three ways for them to travel.[28]

The Growing Girth

The Worldwatch Institute, a Washington, DC based environmental research group released a report in early March 2000, indicating that the world's population is getting more overweight and obese. Problems of overweight and underweight are linked to malnutrition. Well-off minorities in third world countries are growing heavier while the gap between the undernourished widens. The United States and other wealthy countries show a different pattern. The better-educated and wealthier population is getting thinner while the poor are increasing their weight due to an unhealthy diet of cheap and fatty fast foods. In the United States, one in four people are considered obese while 55 percent are overweight. The situation is similar in the United Kingdom, Germany and Russia. If these findings had any bearing on seat configurations, economy seats would be larger than those in premium class.[29]

Physical comfort, especially on long-haul flights, is a primary cause for air traveler discontent. Although the seats vary from airline to airline, the pitch in economy cabins has reportedly declined 2 to 2.5 inches over the last twenty-five years. This reduces the distance between seat rows, effectively moving the passenger in the front closer to the passenger in the row behind. The result is reduced space between one's knees and the seat structure in front, as well as reduced space between the extended seat tray and one's torso. This increases a feeling of physical restriction and heightened discomfort.

High-density seat configurations combined with full loads are the number one factor in heating up passenger frustration, especially on long haul flights. This is particularly the case when cheap fares are a likely guarantee for hours of instant, physical discomfort and a minimum number of cabin crew members to attend to their needs. Being physically ill at ease, being trapped with likewise uncomfortable co-travelers, is a basic recipe for group malcontent.

On a 9-10 hour flight in such crammed conditions, survival and getting your own needs met is basically what counts. Add to this carry-on baggage and no-smoking restrictions, minimum service, unserviceable audio, payment for headsets and bar, seat partners that do not act with the expected decorum of a seasoned airline traveler, and the potential for a tension charged environment is high.

The cumulative effects of physical discomfort under prolonged circumstances are well known to everyone having experienced long haul travel. Tolerance to other environmental irritants diminishes while the urgency to defend the minimal space one has purchased, increases.

Not only is physical discomfort a concern, however medical research specializing in aviation medicine has identified a range of health problems that are of concern to airline workers and that are pertinent to air travelers as well. Some of these issues are addressed in the following section.

Medical Implications

Apart from the obvious discomfort associated with air travel, there are medical implications.[30] The predominant concern is the risk to predisposed individuals suffering from chronic medical conditions that may be aggravated by the cabin pressure. Individuals suffering from respiratory, cardiovascular, cerebrovascular conditions and anemia are well advised to seek medical advice prior to planning any air travel. Developing blood clots or deep-vein thrombosis have also been identified as the result of crammed seating and lack of exercise during flight.

Deep vein thrombosis reportedly caused the death of a 28-year old women on a flight from Sidney, Australia to London.[31] The woman complained of not feeling well during the latter part of the 19,310 km trip and collapsed after arrival. Airlines have come under pressure since it is believed that long periods spent in cramped conditions on the airplanes, known as the 'economy class syndrome' contributed to this condition.

Dr. R. Kraaijenhagen of the university of Amsterdam was subsequently quoted in the media: 'In our study, which is a large study including patients with confirmed thrombosis, all patients with thrombosis had the same percentage of travel as the patients without thrombosis. Traveling is not a risk factor in thrombosis.'[32]

In a study under the Aviation Medical Assistance Act of 1998,[33] the Federal Aviation Administration (FAA) conducted an analysis of cardiac-related problems on 15 airlines for the period of July 1, 1998 to June 30, 1999. A total of 119 cardiac-related incidents were reported, 64 of these resulting in death. The five airlines currently equipped with automatic external defibrillators (AED) used the device in 17 cases and prolonged the life of four passengers. In 40 of the 119 cardiac-related emergencies, no AED was onboard. The FAA proposal[34] announced in May 2000, cites a requirement for AED's, including additional medications and better medical training for cabin crews as part of improving medical care aboard.[35] Clearly, this impacts on the role of cabin crew members, on liability issues and the cost of equipment and training for the airlines.

Hibbert (Public Health Resource Unit, Institute of Health Sciences, University of Oxford) provides some significant insight into why flight might place individuals with medical conditions more at risk of becoming ill and suggests some behavioral changes in others may be attributed to psychological and physiological responses. Hibbert showed a hormone associated with stress (cortisol) increased in anticipation of an event which might be perceived as 'trying' to the individual. In addition, other physiological events occur in response to the flight environment.

Some of the common medical complaints seen on board aircraft, such as fainting or syncope, short of breath or hyperventilation and dizziness can be due to responses to this environment. The cabin air pressure is lower than at ground level which results in a drop in available oxygen. For most, this results in an unnoticed alteration of breathing rate to compensate. For an individual with a cardiovascular or respiratory complaint, however, this response may place them in greater danger of their condition worsening.

The lowered barometric cabin pressure for an altitude of equivalent of 5,000ft-8,000ft (1524m-2438m) results in a drop of partial pressure of oxygen and therefore less oxygen available to the individual. This reduction of oxygen which is usually a fall in lowered barometric pressure from 98mm haemoglobin to 55mm haemoglobin results in less oxygen bound to haemoglobin. ... The response to compensate for such a drop in lowered barometric pressure is to increase the ventilation rate. However, increased ventilation in a compromised individual such as one with a cardiovascular complaint or respiratory disorder can have dire consequences.

The response to this environment is to give up oxygen to the tissues more readily. However, hyperventilating may reduce this effect resulting in the blood clinging on to its oxygen. The lowered oxygen to the tissues, known as hypoxia, can therefore occur very quickly.

The tissues most susceptible to hypoxia are those that demand high levels of oxygen, such as the brain. One of the most common signs of hypoxia is a personality change (Campbell & Bagshaw, 1991). Perhaps a companion may notice this at first, but the signs are not dissimilar to intoxication. The symptoms of hyperventilation are very similar to hypoxia and are more likely to be the cause (Ernsting & King, 1994).

Therefore psychophysiological responses may account for some behavioral changes seen on board aircraft.

The first aid training deals with these conditions, however seldom in the context of behaviour change as a result of these disorders that are aggravated by the physiological environment of the aircraft cabin as well as anticipation of a trying event. Cabin crew members lacking knowledge, coupled with a lack of situational awareness, can increase the risk to a compromised passenger due to incorrectly diagnosing behaviour changes, when timely and correct medical attention would alleviate the condition.

Hibbert provides some helpful hints for cabin crew members to follow, using their observation skills to identifying passengers who may be at risk of developing hypoxia. The list below includes some indications of a person who is distressed or has the potential of becoming ill in-flight. The person may:

- have a distressed look and be very talkative,
- appear to be breathing quickly or breathless when talking,
- use their chest wall to aid breathing,
- punctuate their breathing with sighs,
- not be able to walk a short distance, perhaps from the departure lounge into the aircraft, without becoming breathless,
- look pale or blue (NOTE: colour is not in itself a definite sign but may be an indication of hypoxia in some individuals).

Oxygen levels in the aircraft may aggravate certain medical conditions, such as chronic cardiac or pulmonary conditions. Individuals with these conditions should discuss their travel plans with their physicians to determine if supplemental oxygen should be requested from the airline.

Another common source of discomfort is gas expansion within the gastrointestinal tract. This is attributed to the expansion of gas in lower cabin pressure at higher altitudes. Air travelers are not normally aware of this, however those who have undergone recent surgery for cardiac, pulmonary problems, or eye or brain surgery are well advised to check with their physician before planning to take a trip.

Although first aid training for cabin crews is mandatory and the on-board medical kits provide some good tools for the most common on-board medical emergencies, any individual taking medication should consult their physician in advance of a flight to ensure they have all the information necessary to plan for a trouble free journey.

Air Ventilation Standards

The current level of discussion has not progressed much further. Ventilation standards are not easily obtained nor communicated to the consumer. Reports on the quality of air prior to the smoking ban and subsequent air quality reveal a situation that adds to the sensitivity of the issue. Farrol Kahn, director of the Aviation Health Institute, suggests that aircraft cabins were more thoroughly ventilated when smoking was permitted on board. The average intake of one hundred per cent fresh air was every three minutes. Aviation health experts indicate that twenty cubic feet per minute of fresh air is the optimum amount to ensure comfort. Although procedures for power packs varies somewhat from airline to airline, a reported average of eight to ten cubic feet per minute of fresh and recycled air is available to economy passengers. This procedure increases possible exposure to carbon dioxide, carbon monoxide, neurotoxic hydrocarbons as well as tricresyl phosphates that may result in dizziness, nausea and a feeling of faintness.

Kahn suggests that the overall quality of cabin air was healthier when smoking was permitted because the operating procedures called for greater utilization of the power packs. Airline operating procedures call for an average of forty per cent of re-circulated and sixty per cent of fresh air.

The American Society of Heating, Refrigeration and Air Conditioning Engineers (ASHRAE)[36] investigated allegations on poor aircraft cabin air quality and released its preliminary findings in 1999.[37] Jolanda N. Janczewski,[38] said: 'Environmental parameters, such as temperature or relative humidity, can effect the comfort of the passengers and crew', and suggested a need for development of a testing protocol that could be used in future studies. Her report continued:

> Analyses of the questionnaire responses[39] responses showed that flight attendants tend to cite health-related symptoms, such as headaches, nausea, sore throat and itchy eyes, more often than passengers and are more likely than passengers to relate adverse health symptoms to flying. Except for the flight attendants poor rating of cabin humidity, none of the other air quality factors were rated particularly low or high, including cabin air odor, cabin air quality or lavatory odors. The only symptom passengers were more likely to relate to flying was legs and buttock numbness.

The ASHRAE released a subsequent statement[40] announcing the drafting of a standard to provide acceptable cabin air quality on commercial aircraft. Interestingly, the agreement on a standard so far is defining the minimum ventilation rate by settling on a 5 cubic feet per minute per person (cfm/person) rate of out door air. Alarmists are assured that higher rates are a consideration for different aircraft types.

Obviously, cabin crew member labour associations reacted strongly to the proposed minimum standard.[41] The Airline Division of the Canadian Union of Public Employees (CUPE) organized a conference on October 20, 1999 in Vancouver, Canada for the sole purpose of discussing the long-standing topic of air quality. CUPE's report[42] reviewed events since 1990 showing little encouragement from industry. The Canadian Transportation Safety Board (TSB) however was more supportive. The TSB issued Aviation Advisory[43] alerting Transport Canada to 'a potential aviation safety deficiency (which) ... may affect the ability of cabin crews to carry out their safety-related duties.' The Advisory also called for a study, a recommendation effectively shelved by industry.

If nothing else can be said for certain, airlines continue to be mum on the issue. To the air traveler this should be alarming. If there isn't a problem, why not be more forthcoming, and address cabin crew and growing consumer concerns?

North American attitudes clearly are set on a different path compared with the level of attention by the British government.[44] A parliamentary committee has been formed to investigate the potential link between air rage and the airlines reluctance in communicating health risks associated with air travel. The issue received prominence by the hospitalization of 75-year old Lord Graham of Edmonton following a flight from New Zealand, The House of Lords Science Committee will conduct its inquiry over a six-month period. Airline executives will have to answer to accusations that their operating procedures concerning cabin air ventilation, and crammed seating is driven by greed at the detriment of the health of their customers.

The Question of Ozone Exposure

Cabin crew member labour associations raised the question of cabin air quality since the 1970's coinciding with the introduction of jet aircraft. At first the focus was on ozone levels, especially on long-haul flights operating over the North Pole. The issue surfaced in 1985 with one of the major Canadian carriers encouraging the reduction of air conditioning pack use in order to save fuel cost. Analysts estimate savings are up to six per cent per year or $60,000 per plane or approximately $1 per passenger depending on the cost of fuel (Pieren, 1997).

Incident reports of cabin crew members and passenger illness ranged from fainting to dizziness, and nausea. Arguments focused on the better airflow in the cockpit at seven to 12 liter per second per crew member compared to the air flow in the passenger cabins.

Findings, predominantly on the bases of analyzing incident reports, were considered by the medical department too inconclusive for establishing a clear link between ozone exposure and the use of the power packs and reported illnesses. The issue of relating illness to one singular cause is difficult to establish since these types of complaints can be triggered by multiple factors unrelated to exposure to cabin air. The result is a simmering debate now well into its fifteenth year.

The non-disclosure of readily available information on the effects of ozone exposure points to an issue of organizational ethics, misplaced trust and complacency. As one member of the In-Flight Service management team involved in the discussions recalls: 'I expected the medical representative to inform us on research conducted in this area. Surely as experts they would provide us with a synopsis of findings to date. None of this occurred.'

A recent search on the net produced sixteen articles published on ozone exposure and its effects in *Aviation, Space and Environmental Medicine* between 1962 and 1986 alone. Topics range from acute effects to long-term ozone exposure, cardiopulmonary function, toxicity hazards, effects on visual parameters and psychomotor performance during ozone exposure.

Altered Senses

Lesser-known and published effects of air travel are the changes to our sense of taste and smell. This is attributed to the reduction of oxygen when flying at high altitudes. Airlines more concerned with these effects create their menus to address this fact. In addition to planning food that holds moisture to off-set the dry cabin air, food allergies, triggered e.g. by seafood, are also taken into account.

On long-haul flights with an ethnically mixed customer base, e.g. Western and Asian, established and thoughtful airlines take much care to accommodate customer needs and preferences. On a thirteen-hour flight, customers can expect two hot meals and a snack, much the same food intake one would expect to have when not traveling by air.

The choice of special meals for health and religious reasons has proliferated significantly with global air travel. International air carriers demonstrate a keen appreciation of their varied customers and their dietary needs. On average, there are approximately 17 to 20 different types available, a clear indication of the time, effort and resources dedicated to satisfy these needs.

Summary

With changes in public conduct, the changed demographics and greater number of air travelers on the move, more openly voiced discontent with the industry, airports attracting more diversified crowds and yet bigger aircraft close to the manufacturing phase, the impact on passenger-airline relations is measurable and far-reaching.

As relationships become more tenuous, a new vision for the role of airline workers is a must. Organizations that have failed to take steps to address this issue for the last ten years or more are well advised to remedy this situation. As an example, one international carrier completed such a review last in 1989. Performance standards and measurements have not been adjusted since.

Although the focus on training has been modified to a certain extent to deal with the issue of passenger misconduct, the brunt of these changes affects mostly procedures dealing with serious incidents. Regulations concerning the management of medical onboard emergencies have been in place for some time and are subject to regulatory review and refinement. The need for a more holistic approach might prove to be more effective when aircraft manufacturers and airlines develop their plans for new cabin interiors.

The physical environment of travel by air is significantly different from any other transportation mode, be it by sea, train, bus or car. The impact on the air travelers' well being is not well-understood and subject to individual physiological conditions, including psychological pre-disposition. With the rich body of research available on the topic of Human Factors and its physiological affects, aviation leaders can easily draw parallels to the plight of the air traveler. Responsible decisions concerning allocation and quality of seating configurations, catering and training of staff are marketable and can be used as competitive levers.

Legislation – a Change Engine

Air travel is accessible to a wide range of international clientele, of a broad social spectrum in a transformed society with diverse expectations and values. Education, financial means, social status, and cultural background and beliefs are no guarantee for adequate socialized behavior in the aircraft cabin.

The liberalization of human rights legislation has resulted in the dismantling of certain screening safeguards, which previously gave airlines some control to restrict travel under certain conditions.

As an example, the Canadian Human Rights Act, passed by Parliament in 1977, specifically protects anyone living in Canada against discrimination in or by airlines. Grounds for action against an employer or provider of service that fall under federal jurisdiction cover a wide range. Discrimination on the basis of race, colour, national or ethnic origin, religion, age, sex (including pregnancy and childbearing), marital status, family status, physical or mental disability (including dependence on alcohol or drugs), personal criminal conviction, and sexual orientation, are all specifically mentioned. Despite the benefit of this legislation biases against people who are perceived to be different are acted out daily, including in the airline industry.

Legislation shapes society and shapes business along the way. The airlines are no exception. Legislation ensuring accessible transportation for travelers with disabilities had a profound effect on airline operations. Additional cost to progressively remove barriers and to train employees who serve these customers is an ongoing concern for the aviation community.

Acknowledging Special Needs

British Airways does not only contemplate improvements for business travelers alone. The airline wants to make life easier for parents traveling with infants and has ordered 1,000 luxury sky-cots that can be attached to bulkheads in the aircraft, allowing babies to fly lying down more comfortably than in the previous generation of sky-cots.[45]

People with disabilities are also getting help. Air Canada was one of the first air carriers in the world to introduce on-board wheelchairs for the convenience of the disabled traveler. Unfortunately, little use is made of this feature for a number of reasons, not the least for a lack of promotion by cabin crews.

WestJet Airlines, a low-fare up-start operator in Western Canada, has introduced a new product to assist in the transfer of wheelchair users to and from the Washington chair to their assigned seats. Comments from Sharlene Taylor Manager, Information and Referral at the Independent Living Resource Centre in Calgary, Alberta, capture her delight:

> As a first time experience, I was able to slide across to the window seat, which I have never done in all my years of travel. I would like to see, at some time, that all travel services have access to this kind of equipment for the safety of anyone assisting people with disabilities in transferring while traveling. It made my transfer much more dignified.

The Canada Transportation Act

One example of such legislation is *The Canada Transportation Act*. It underlines the Government of Canada's commitment to equitable access to transportation services by all travelers including persons with disabilities. Under the Act, the Canadian Transportation Agency has the power to remove unnecessary or unjustified obstacles from Canada's transportation network that includes air carriers and airports, passenger rail carriers and stations, and inter-provincial ferry services and their terminals. To achieve the goal of accessible transportation, the Agency consults extensively with the industry, travelers with disabilities, associations and people interested in these issues.

In the United States a similar act, Americans With Disabilities Act (ADA) provides comprehensive civil rights protection to individuals with disabilities in the areas of employment, public accommodation, State and local government services, transportation, and telecommunications. Details concerning this act can be found on the FAA web site.

How the Canadian Transportation Agency Works

Through the Accessible Transportation Directorate, the Agency's program of accessible transportation includes three main activities: Complaint Resolution, regulations/guidelines, and monitoring and liaison.

Complaint Resolution

A person with a disability encountering an obstacle while traveling may choose to file a complaint with the Agency. The Agency must then consult those involved and make a decision within 120 days unless the parties agree to an extension. If the Agency determines the problem is an "undue obstacle", it may order corrective action. For example, the Agency may order the removal of the obstacle or the payment of compensation. In some cases, both remedies are taken.

In many instances, a traveler with a disability may resolve a problem by directly contacting the transportation service provider. However, if the problem is not settled to the satisfaction of all parties, the Agency can help resolve the dispute.

Regulations/Guidelines

The Agency has the power to develop regulations or guidelines to eliminate unnecessary or unjustified barriers to the transportation of persons with disabilities. These barriers can cover fares, conditions of travel, the training of staff, signage and the way information is provided, as well as the design, construction or modification of aircraft, rail cars, ferries and terminals.

Regulations apply to Canadian carriers operating domestic services using an aircraft with 30 or more passenger seats. Specific mention is made for assistance with registration at the check- in counter, assistance in proceeding to the boarding area, assistance in boarding and deplaning, including transfer and lifting from the mobility aid to the passenger seat, and assistance in stowing and retrieving checked baggage.

The regulation further specifies what services can be expected on board the aircraft. Crew members are required to assist with putting away and retrieving carry-on baggage, in general assist, other than carrying, to move the traveler with a disability to and from the washroom, serve special meals, and help with meals such as opening packages, identifying items and cutting food.

The Agency implemented a set of regulations, *Personnel Training for the Assistance of Persons with Disabilities Regulations*, outlining specific training requirements for the personnel of most airlines and airport operators to meet the special transportation needs.

Monitoring and Liaison

The Agency places a high priority on information sharing and creating awareness of the problems experienced by travelers with disabilities. It therefore consults regularly with groups and associations representing persons with disabilities, government departments and representatives from the transportation industry. In addition, the Agency studies issues and conducts surveys relating to accessibility in the transportation industry.

The Agency's *Guide for Persons with Disabilities, Taking Charge of the Air Travel Experience* is an excellent resource. It offers helpful tips and a check list from planning a trip, to making reservations, to seat selection, through the terminal, boarding the aircraft right to the end of the trip. An equally useful tool is the guide *Making Specific Needs Known*, designed to help travelers with disabilities plan and prepare their trip by air within Canada. The Agency's web site is comprehensive, well structured and informative. It reflects its commitment to its main users with practical, well thought out material and sensitivity.

Air Travelers with Disabilities

Air travelers with visible and invisible disabilities are still facing a multitude of obstacles. These range from physical to attitudinal barriers when dealing with airline staff. Rick Goodfellow, Executive Director of the Independent Living Resource Centre of Calgary, speaks from experience:

I was traveling to Toronto with a couple of colleagues to attend a meeting. On arrival in Toronto, while waiting to be taken off the airplane, the gate agent came after all other passengers had left, and advised me they were waiting for an aisle chair (a wheelchair designed for transfer in the passenger cabin) to be brought on board. My partner about to head up to the gate to make a phone call, acknowledges: We know, we are waiting for a Washington. (A 'Washington' is an airline term for the special transfer wheelchair).

After a while, this big fellow comes along with the Washington chair and moves me onto it. Although there is a seat belt to secure me, he chooses not to use it. At the doorsill of the aircraft he tips the chair back and pushes it over the gap between the doorsill and the bridge leading to the terminal. Bridges are seldom perfectly aligned with the aircraft door, and the wheelchair's casters are small enough to get stuck in the gap. The agent pushes me vigorously over the gap while I am feeling every bump and trying to balance on the chair. He then rests the wheelchair facing the aircraft door in the bridge while I am bracing myself in order not to slide off the chair. My partner coming off the aircraft sees this and asks the agent: 'Shouldn't he be strapped in? The agent responds: ' I'm a pretty big guy, and I think he is alright. I'm not worried about him, but if you are worried about it, next time, you do it (strap him in).' My partner insists: ' I am worried about it, and I believe it is regulation to strap him in.' The agent comes back with: 'I'm telling you, I'm not worried.'

Meanwhile, I am still bracing myself and control my growing agitation. I feel vulnerable at the hands of this agent and believe that anything I want to say at that moment won't ease the situation. Just then, my other partner who had gone ahead of us earlier to make a call in the gate area returns and witnesses this last confrontation. He asks the agent for his name who refuses to provide it. The first agent returns without my chair. My colleague tells him, that we are still waiting for it. 'Oh, I sent it on to Washington,' and hastily departs.

Stranded, with no apology from either agent, the wait continues, and our anxieties and frustration rise. Later, on the way to the baggage area, my partner stops three times to ask for a Supervisor. No one responds positively. The third agent tells us that the Supervisors are all in a meeting and cannot be disturbed. At that time, my partner, Chairman of the Advisory Committee for Accessible Transportation, (a group directly reporting to the Minister) identifies himself as such and demands to speak to a manager. That's when we finally got action. As an ordinary consumer, we got none.

Observations

This anecdote leads to the following observations:

1. Goodfellow did not communicate with any of the agents directly.
2. Third party communication occurred only between his colleagues and the airline staff.
3. The mention of a rule having been broken did not elicit a positive response.
4. The agent, finding himself in a situation of being criticized, took the risk of causing potential injury to Goodfellow and liability for his company.
5. The use of airline jargon ('Washington' for the aisle chair) resulted in an error that further exasperated the situation.
6. Previously uninvolved airline staff showed no willingness to address and act upon the complaint.
7. The eventual threat of the passenger's professional status prompted the agent to bend the rule of not interrupting the meeting.

Analysis

The agent's behaviour in refusing to buckle up Goodfellow is typical of poor performers who take great risks in bending rules. The agent needed to demonstrate control of the situation at all cost. The overconfidence of the agent who related the term 'Washington' to a connection destination may have been caused by a number of factors:

1. The agent was uncomfortable with the conflict situation in progress and wanted to leave.
2. The need to deal with the personal wheelchair provided a reason to act.

3. The level of discomfort with the situation prevented him to clarify an important piece of information.
4. The agent simply felt he had the correct information and decided to act.
5. The agent was under time pressure and decided to act.

The airline staff's refusal to deal with the irate passengers may have been prompted by a number of factors. Complacency, the lack of a well-defined and understood company compensation policy, an uncaring attitude – 'they get over it,' unwillingness to take ownership for a problem they did not cause 'themselves, are some of the possible causes.

Only when one of the passengers wielded his special social and professional status as a lightening rod, did the last agent make the correct decision. She broke the internal rule of not interrupting a meeting, and called a Supervisor. She recognized the potential for the serious ramifications to the company's public image and punitive measures via the Minister's office.

Summary

In this example, short-term satisfaction was achieved. The question remains what if anything was done to change the behaviour and problem solving skills of the agents involved in this incident? Most likely nothing was done. Operational and other management priorities usually overshadow such incidents. The manager at best praised the Supervisor for making amends, while a follow-up investigation with the staff was deemed unnecessary.

Uneasy Relations

Travelers with disabilities are not favoured customers. Advocacy has forced legislation, and legislation forces the airlines to provide certain standards of service. Assisting air travelers takes more time compared to the 'low maintenance' average passenger. As on-time performance objectives drive the operation, the boarding of a passenger with disability causes a wrinkle in the tight time frames afforded to the airline staff. It follows, that shortcuts are taken, resulting in a less than dignified experience for the disabled.

It is estimated that the annual incidence of spinal cord injury (SCI), not including those who die at the scene of the accident, is approximately 40 cases per million population in the U.S., or approximately 10,000 new cases each year. The number of people in the United States who are alive today and who have SCI has been estimated to be between 721 and 906 per million-population. This corresponds to between 183,000 and 230,000 persons. The growing awareness of the potential revenue expected from this group alone is cause for airlines to make greater efforts in accommodating their needs.

The Lack of Documented Complaints

The disabled community has many complaints. Although legislation in highly industrialized countries has improved accessibility to public transportation and buildings, daily experiences of insensitive individuals in the service business are common. Why then, are complaint statistics so low that they do not prompt any concern?

The reasons most frequently cited are a fear of disclosure resulting in systematic backlash from those one depends on. This has been voiced both as a group and as an individual concern. Other reasons mentioned are the desire to distance oneself from an unpleasant event, the skill and time it takes to write complaints, and lack of trust that one's efforts bring positive results. Time involved in writing complaints is a big factor. Most daily functions taken for granted by able people, take considerable more time and effort to accomplish by a disabled person.

'Air Rage' – Not for the Visibly Disabled

Male and female passengers from diverse national, social, economic and educational background have reportedly engaged in incidents of misconduct except one group of air travelers, the visibly disabled. The question is why not? As much as we need to understand the reasons and contributing factors leading to passenger misconduct, much can be learned from examining the reasons for As illustrated in examples of decisions rendered by the Canadian Transportation Agency in 1999 and 2000, multiple service failure, physical and travel restrictions, including unavailability of personal mobility aids are certainly serious antecedents to potential passenger misconduct.

What is it that distinguishes the visibly disabled from the invisibly disabled engaged in situations of excessive misconduct? It is the utter physical dependency, the inability to exert the most basic physical threat to another person and incur bodily harm. Deprived of their mobility aids, the dependency is complete, the surrender of the most important means of independence absolute.

Canadian Transportation Agency Decisions

The following experience leading to an official complaint allows the reader to gain an appreciation of the added sources of major frustration and inconvenience from which able-bodied air travelers are spared. A sample decision is provided in this chapter (with a second one in *Appendix D*) illustrating the scope of complaints as well as the scope of issues. The decisions reflect the limited powers of the CTA, especially in regards to compensation. Airline contracts of carriage are not making special allowances for compensating the severely inconvenienced disabled traveler. This gives the disabled air traveler a greater disadvantage compared with the rest of the passengers.

Decision No. 563-At-A-1999

Issue: The issue to be addressed is whether carrier X refusal to transport Ms. P. from Montego Bay to Toronto on December 27, 1997 due to the unavailability of an oxygen kit constituted an undue obstacle to her mobility, and, if so, what corrective measures should be taken.

Facts: During Ms. P's stay in Jamaica, she experienced a stroke. When Air Canada was informed of her situation, its Medical Officer informed her that she could travel two weeks following her stroke with a medical attendant and a supply of oxygen. After three weeks, she would have been able to travel without an attendant or an oxygen kit, but would still need medical authorization. Ms. P. chose to travel after two weeks.

The final arrangements were made for her to return to Toronto on December 27, 1997 on carrier X Flight No. 982, on which the carrier loaded a Medipack oxygen kit that had been sent from Canada. However, when Ms. P. and her doctor presented themselves for the flight, the agent could not locate the oxygen kit. Carrier X therefore did not permit them to board the aircraft and made arrangements to have them travel the next day. The oxygen kit was located and made available, which allowed Ms. P. and her doctor to travel on December 28, 1997.

Carrier X's procedures at the time of travel did not specifically identify where the Medipack kits should be stored. However, on June 2, 1998, the carrier amended its procedures to ensure that all Medipack kits are stored in a designated overhead bin so they are easily located.

Positions Of The Parties: Carrier X apologized to Ms. Powell for the involuntary postponement of her return. It adds that it is confident that the new policy will prevent similar incidents in the future.

As a result of the failure to provide Ms. P. with an oxygen kit on December 27, 1997, she is seeking compensation for the services of:

- the doctor who accompanied her, which she submits carrier X had promised to pay;
- the caregiver who attended to her needs on December 27 and 28, 1997;
- the chauffeur who transported her between the airport and the hotel; and for long distance telephone calls; and
- the stress and duress she experienced due to carrier X's failure to provide the oxygen kit.

Although there were discrepancies in the costs Ms. P. identified in her documentation, Ms. P. submits that the expenses included $2,000.00 for the doctor, $300.00 for the caregiver, $150.00 for the chauffeur, and $356.18 for the telephone calls.

Carrier X states that, at the time of this incident, it had agreed to pay $2,000.00 compensation for the doctor who accompanied Ms. P. which carrier X states it would do when it received the doctor's original invoice. Carrier X indicates that it informed Ms. P.'s daughter that it would consider the expenses she incurred as a result of its inability to accommodate her transportation on December 27, 1997, provided that they are supported by appropriate official documentation.

In response, carrier X provided Ms. P. with a cheque in her doctor's name in the amount of $2,000.00. It also indicated that, as a gesture of goodwill and concern, it included a cheque to Ms. P. for $356.18 to cover her telephone expenses and a travel voucher for $150.00 to cover her ground transportation. Ms. P. indicates however that she is not satisfied with the offer as it does not include payment for her caregiver for the additional services required on December 27 and 28.

Analysis and Findings: In making its findings, the Agency has reviewed all the material submitted by the parties during the pleadings.

The Agency finds that carrier X's denied boarding of Ms. P. due to the carrier's inability to locate the oxygen kit on the December 27, 1997 flight constituted an obstacle to Ms. P. mobility in that it delayed her return home by one day and required her to extend the services of her doctor and caregiver by one day. It is the Agency's opinion that measures should have been in place to ensure that such a crucial piece of equipment deemed necessary for the transportation of a person with a disability was indeed available at the time of the flight. The Agency finds the obstacle to be undue as it could have easily been avoided had carrier X's procedures thoroughly addressed all aspects of the provision of an oxygen kit, including where its agents could locate it in the aircraft.

With respect to Ms. P.'s request for compensation, subsection 172(3) of the CTA limits the Agency to direct the payment of a compensation to a person with a disability for the costs incurred as a result of an undue obstacle. In this respect, the Agency finds that the expenses incurred by Ms. P. for the additional service of the doctor as a result of the delayed departure, as well as the additional costs of the caregiver, the telephone calls and the ground transportation constitute expenses incurred as a result of the undue obstacle to her mobility.

The Agency does not have jurisdiction to award damages for stress and duress.

The Agency notes that carrier X has, as previously agreed, issued a cheque in the amount of $2,000.00 to the doctor.

With respect to the additional expenses for the ground transportation and telephone calls, although the Agency has not set the amounts of the eligible expenses, carrier X has paid the full amount claimed by Ms. P. for the telephone calls and provided a travel voucher for the full amount claimed for the ground transportation.

The Agency notes that the offer made by carrier X with respect to telephone and ground transportation expenses was made as a gesture of goodwill and concern. However, the Agency considers the reimbursement of eligible expenses incurred as a result of an undue obstacle to be an entitlement to the person with a disability.

The Agency notes that Ms. P. has accepted the compensation offered for these expenses. In light of this, the Agency finds that no further action is necessary with respect to the expenses for the doctor, the ground transportation and the telephone calls. However, as carrier X has not addressed the cost of the caregiver, which has been deemed an eligible cost, carrier X is required to reimburse Ms. P. for this expense.

Conclusion: Based on the above findings, the Agency determines that carrier X's denied boarding of Ms. P. due to the unavailability of the required oxygen kit constituted an undue obstacle to her mobility. Carrier X has since amended its procedures to ensure that Medipack kits are stored in a designated overhead bin so that they are easily located. It is the Agency's opinion that this change in procedures should help to prevent a recurrence of similar situations in the future.
With respect to financial compensation, the Agency, pursuant to subsection 172(3) of the CTA, directs carrier X to reimburse Ms. P. for the additional cost incurred for the services required by Ms. P., which were provided by the caregiver on December 27 and 28, 1997 due to the delayed departure.
In the event that Ms. P. and carrier X do not reach an agreement on the above eligible cost, they may bring this matter back to the Agency for its determination of cost and issuance of an order to that effect.

Summary

The decision the Agency has rendered in this example and in the sample cases provided in *Appendix A,* reflect its restrictive powers and a preference to use collaboration with the airline industry rather than following the route of litigation. The emphasis is on reimbursement of expenses incurred as a result of the failure to comply with current regulatory requirements. Any reference to the potential for changes in the air carrier contracts is avoided.

Issuing bulletins is a quasi management placebo to indicate action has been taken, but the real issue is missed. Bulletins indicate to outsiders of the organization that policies and procedures are in place to deal with a variety of business related issues. They are no confirmation that they are being practiced. Unless management is prepared to implement an organizational structure and means to monitor employee performance effectively, the issuance of bulletins is a waste. The Agency itself is restricted in enforcing its regulation due to the limited number of enforcement officers in its employ.

Since the relationship between Canadian air carriers and travel agents has deteriorated over the last couple of years due to severe cut backs of commission rates. Travel agents' motivation in carrying out their responsibilities under regulation SOR/94-42 is undermined. To illustrate this situation, in October 1999, one major Canadian air carrier reduced the commissions from eight per cent to five. This is a further reduction from rates established in 1997, when carriers reduced the rates from ten to eight percent. These dramatic changes affect the way business travel agencies conduct business with their customers, including adding fees for services that customers used to obtain for free.

The question of service received under the provisions of the Tariff Act and regulation SOR/94-42 versus compensation for disservice rendered is open for debate. The airline traveler is treated quite differently from any other consumer having received a faulty product. In the case of having paid with good money for a faulty product, the product is replaced by a reputable business, at no cost to the customer. In the case of the air traveler, the experience of a significant service failure is treated with surprising superficiality by the airline industry. How do you replace hours and days, at times weeks of being inconvenienced by an airline, simply because it does not manage to ensure the performance of its employees?

A closer examination of compensation policies and service recovery programs reveals airlines are severely lagging behind other service industries excelling in quality service to their customers. Giving bonus air miles or travel vouchers are inadequate. They have little value since redemption is increasingly difficult and subject to a high number of conditions, a source of further aggravation rather than a true gesture of making amends.

Challenges

The diversity of today's traveler presents enormous challenges for the airline industry. Airlines operate on the principles of conformity when the customer is looking for increased freedom and control. The air traveler is looking for service on a human level, face to face communication when all goes wrong, while the airlines are getting more and more production oriented. Automated processes are here to stay and will be expanding at a quick pace over the next ten years. A communication and information strategy needs to be formulated to addresses the biggest complaint of air travelers in distress: 'Nobody told us anything.' The frustration level of the air traveler is intense.

In the age of information technology and the ability to freely access information, extended silence from the airline representatives in situations of operational disruptions such as weather or maintenance, is unacceptable. Person-to-person communication with people under duress is quite different than disseminating information electronically. Efficient and customer centered decision making and problem solving under high pressure are increasingly required to deal with operational problems, no matter what the cause. Comunication skills must continue to be developed and remain an important corner stone of the airline business.

Airlines are at the leading edge of technology using automated systems to streamline many administrative functions, achieving greater consistency and eliminating a host of human errors. This results in fewer jobs and greater cost efficiency. The downside of this trend is the diminishing quality of personalized service the traveler experiences, and the airlines' diminishing sensitivity in fulfilling their responsibility in providing meaningful and timely information person-to-person.

The Physical Benefit of Communication

In business as well as in personal relations, the virtues of communication play a major part in contributing to satisfaction and success. Less publicized are the effects of giving information on the physiology of stress. Hibbert provides us with the following insight:

> Studies have demonstrated the importance of giving information in order to alter physiological responses and improve the outcome (of stress). For example, it has been demonstrated that patients who had been psychologically prepared for surgery by being informed of what to expect secreted lower levels of corticosteroids in their urine compared to patients who received less information (Boore, 1978). In another study, an anticipated endurance exercise of a 20 kilometer walk with a full backpack for a group of Israeli soldiers resulted in increased levels of cortisol. There were three sub-groups among the soldiers and the difference between the sub-groups was the information given to the soldiers about the event. Those informed precisely and frequently about the duration of the exercise fared better than those who were misinformed. Interestingly the soldiers who performed worst in that they recorded higher levels of cortisol, greater exhaustion and a perceived inability to cope, were not those uninformed but those who received no information at all (cited by Ursin& Olff, 1993). ... Further work by Levine and Coover (1976) (cited by Ursin &Olff, 1991) showed that feedback resulted in lowering the circulating corticosteroid levels.

This knowledge of physiological responses to information is being increasingly applied in the health care sector that has seen a major change in the doctor-patient relationship. Courses designed to help people with a fear of flying and offered by some airlines such as British Airways, Lufthansa and Air Canada use continuous information during a flight to reduce the physiological responses to stress.

Airline policies concerning public announcements of the 1960s and 1970s at Air Canada placed much greater emphasis on commentary in-flight by the flight crew and the In-charge cabin crew members. With greater public sensitization to traveling by air, some of this in-flight chatter became too redundant and interruptive for the more frequent flyer. Bilingual requirements and policies concerning route languages doubled, and multiplied depending on the passenger demographics of the day demanding further translation. What had set out to be a stress reducing exercise now turned into a stress factor for those passengers who had already heard the announcement in the language of their choice.

Regulatory driven safety announcements are bothersome to the seasoned air traveler and a source of irritation. Air travelers at large are well versed and willing to cooperating with regular safety preparations. The repetitiveness of these announcements wear the listener's nerves thin over time.

In summary, the giving of information is not 'just a polite thing to do'. As the reader has seen, it can have very positive effects on the physiological outcome of stressful situations. No doubt, industry's silence on air quality issues and its reported effects on comfort if not health, risks the spreading of misinformation.

Sensitivity to media coverage reflects industry's deep concern to protect its reputation. The lack of conclusive scientific evidence surrounding health issues attributed to the aircraft cabin environment add to uncertainties, however airlines cannot avoid responding to growing confrontational pressures from influential interest groups much longer.

Information must be designed to meet the audience's needs, be timely and accurate for it to be effective. With marketing taking advantage of the latest technology and psychological persuasive methods to deliver their sales messages, the poor the level of communication between the airlines and their customers at the operational level is hardly excusable.

Monetary Value of Communication

Being an experienced air traveler does not mean greater tolerance in accepting the irritants associated with air travel. The following samples of complaints from just one frequent flyer illustrate that providing timely and factual information to the customer when things go wrong is very important in keeping their good will. The first letter is addressed to Continental Airlines and copied to the President, Mr. Gordon Bethune.

June 30, 1999

US Department of Transportation Consumer Complaint
400 Seventh Street SW
Room 4107
Washington, DC 20590

Dear Sir or Madam,

Please find attached two letters sent to Continental Airlines regarding the difficulties Mr. Sheehy has experienced recently.

We are sorry to say that these are not the only occurrences because there have been many, but these were the ones we highlighted.

If you require additional information, please don't hesitate to contact me at the numbers below.

Sincerely,

Tuesday, March 16, 1999

Dear Customer Care:

We would like an explanation as to why on Monday, March 15 Flight 4282 departing Savannah, GA at 5:20pm arriving Newark at 7:35pm, that Continental Airlines made no attempt to notify passenger Mr. Barry Sheehy that Flight #4023 departing Newark on route to Savannah, GA was *already* two hours behind schedule. If Mr. Sheehy had been notified, he of course could have left later for the airport or had the choice to make alternative arrangements.

Five calls were made randomly throughout the day to Continental Airlines beginning at 9:30am and the last call being made at 3:50pm all ensuring that Flight 4282 was on time for departure. Upon checkin no one advised Mr. Sheehy that there was any difficulty with the flight's departure time. At what should have been boarding time, an announcement was made indicating Flight 4282 was being delayed 2 hours. At the new delayed departure boarding time, an announcement was made canceling the entire Flight. No provisions were made to secure Mr. Sheehy on your last flight out of Savannah departing at 8:15pm arriving Newark.

No extra staff was put in place to help customers get information or reschedule flights etc It appears that there was a total lack of appreciation for the customer and a complete breakdown in communication between airline and staff and staff and customer. Customers either in the airport or calling on the phone were being given inaccurate information. Continental Airlines had to have had current and dependable information with which to update their staff and customers prior to the cancellation.

Please advise us as to how you plan to compensate Mr. Sheehy for this exasperating experience. Also and most importantly, how you plan to reassure us that we can depend on the information being given by your staff both in the airport as well as on the phone.

We look forward to your reply.

Yours truly,

V.W

Another letter is dated June 30, 1999, again addressed to the President of Continental Airlines, and reveals the airline is unresponsive, and recovery has failed:

Dear Mr. Bethune:

On Monday, June 28, Mr. Barry Sheehy was scheduled to be a passenger on your Flight 4020 departing Savannah at 12:30pm but it was cancelled. Mr. Sheehy was never notified of the cancellation, nor was his office. Calling to confirm flights prior to take-off has become an unreliable method to receive information. My experience has been that almost up to departure time your representatives advise that the flight is departing on time, which in many cases is not correct. I therefore have learned that I cannot depend on this information.

I have attached a letter of complaint, which I forward to you, on March 16, 1999. We never received a response from your office. Mr. Sheehy flies routinely from Savannah to New York, Monday to Friday, Continental would be his airline of choice but the unreliability of flights makes it difficult.

I would be interested to know how you can reassure us that we can depend on the information being given by your staff. We look forward to your reply.

Yours truly,

V.W.

The tone of these letters is objective, stating the facts, and simply inquiring about the reasons for a lack of communication. They are a testimonial of the frustrations endured while trying to do business with the airline. The unfortunate part of handing good money over to the airline in advance of experiencing the service is the high risk involved for the consumer.

Other situations involving children traveling on their own, causes at times great anxieties, for parents, especially when things go wrong. One such incident involved the cancellation of a flight out of Newark. The airline refused to re-book two girls, aged 15 and 14 on the next flight. They were prepared to re-book the 15 year old, but not the 14 year old without so much as an explanation or concern for the ramifications on either these minors or their parents.

Understandably, during irregular operations airline staff is at the brunt of angry and frustrated passengers, however re-booking procedures must address the carriage of under-aged air travelers with a concern for their well-being. This is a matter of social responsibility and liability if things go wrong.

Air Traveler – Protect Yourself

As an air traveler, there are certain guidelines for their own protection that range from health to security to dress code tips. With the widely accessibility of electronic information systems, air travelers are in an excellent position to educate themselves on how best to interact with the aviation system. Government sites are comprehensive and offer factual information that assists the potential air traveler in preparing for a less stressful experience, provided safety and security procedures are voluntarily implemented, and common sense guidelines are followed. Other sites such as 'AirSafe' offer practical advice using a more conversational style. We already referred to airline web sites as a good source for finding out what specific policies and products the consumer can expect.

Air travelers suffering symptoms related to cabin air quality are encouraged to:

- Complain to the cabin crew;
- Follow-up to the airline with a written complaint;
- Lodge a report with a consumer action group;
- Lodge a complaint with civil aviation authorities.

Consumer agencies are also offering comprehensive information for prospective travels, including special sections aimed at women with the aim to instill norms of conduct that guard women from becoming unwittingly prey or victims. The range of common sense hints is wide-ranging and recommended to reading, especially for the novice traveler.

The airline industry for example, issues dress codes for their non-revenue passengers for good reason, inoffensiveness and decorum make good sense. If these guidelines are not followed, travel can be denied and in worst-case scenarios privileges revoked.

The practicality of a dress code is based on comfortable and more conservative attire. For one, should there be an emergency, clothing acts as protection in case of fire and certainly during evacuation. Zooming down an evacuation slide in shorts, hot pants or mini skirts will leave major burns. If this rational is insufficient, considerations for one's own personal safety, especially minors, should be examined.

Regardless of the reason for travel, be it business, leisure or personal, a basic standard would reduce embarrassment, unwanted attention or sexual advances during a trip. General preparation for safe footwear, coverage of legs and upper body are recommended. In a public place, in very close quarters and physical contact with strangers from all walks of life and countries, personal hygiene and discretionary attire with consideration for other people's sensitivities goes a long way to making a trip more pleasant for everyone. The proliferation of hints for civility in the air has become a source of both amusement and common sense. 'Miss Manners' opinions are clear: stress is no excuse for offensive behaviour and personal habits. She promotes the view that people have a choice. Her advice is simple to make the right one for everyone's benefit.[46]

Rudeness as a means to achieve satisfaction is a sure way of at least encounter tacit resistance. As a power base for social interaction, it is by its nature destructive. Engaging the airline worker in solving the problem rather than spitting insults is infinitely preferable for everyone's sake.

Training

Major air carriers, especially prior to expanding their route system provide special training addressing cultural differences of the new clientele. These efforts are commendable, providing that some form of recurrent cultural difference training is also integrated into initial cabin crew and recurrent training. A one-time deal is insufficient, due to the need to sensitize crews periodically to cultural differences placing different demands on the type of social interaction on board, especially when dealing with the enforcement of rules and regulations. With the acceleration on production in the air industry, demands on better processes, including training are very much needed to affect greater social peace in the passenger cabin.

Third Party Training

The increased practice of outsourcing training in the aviation industry, particularly of flight crew training on simulators, is also becoming a practice in outsourcing training in conflict resolution and combating air rage. There are a number of problems associated with this approach.

First, there is limited input by the air carrier on the specific priorities evidenced in their own operations. Training becomes generic from that vantage point, and misses out on the opportunities to deal with issues that may be unique to this carrier. Well-developed internal statistics regarding the number, type and severity of incidents provide a more meaningful base to examine policies and procedures, including training needs of that particular workforce. Second, joint training efforts are further complicated under these conditions and unlikely to be tailored to the specific needs of both groups. Third, this type of training overlaps with the development of work skills in the area of handling passenger complaints and compliance issues. We therefore support a more integrated approach in all training phases of both the flight and cabin crew.

The same set of challenges exist for airlines that have decided to sever organizational ties with their training centers and turn them into independent profit centers. Proponents of this move tout the advantages to sell their expertise and generate revenue to offset the high capital cost associated with training in the aviation industry. Lufthansa and British Airways are two of a number of organizations that have undertaken this move.

Prioritizing Air Traveler Relations

Figure 3.1 provides an overview of issues that must be addressed by airlines in the development of a preventive strategy for achieving social peace in the passenger cabin. It features five main components: the airline organization, pre-conditions, mitigating factors, and customer prioritization issues relative to the customer base.

A market analysis and subsequent marketing plans have to take into account the demographics of the customer base, including their nationality, gender, cultural uniqueness, values and experience in air travel. There is no doubt a greater need to include pertinent aspects of the rich body of research in the social sciences, psychology and anthropology to complement traditional marketing models.

As a service provider, questions of responsibilities regarding passenger education cannot be avoided. The traditional approach to fixing a lack of public awareness on-board through added public address announcements are no longer meeting their target. They are usually too late and just add to a growing number of announcements required by law. Add to this the number of translations to meet route language requirements, and the likelihood of verbal saturation sets in quickly.

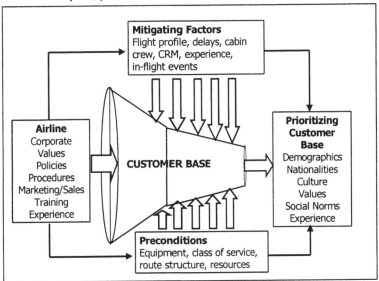

Figure 3.1   Prioritization Model

Airlines would benefit from taking on the task of better passenger education, not only in terms of on-board safety but also from a competitive point of view. Misleading or confusing information will do little to improve the already stressed relationship between the airlines and their customers. Education does not have to be patronizing, but made interesting and appealing. Opportunities to develop partnership with educational media experts could lead to a worthwhile and rewarding venture for all involved.

On-board people management is very important. Airlines recognizing this fact will no longer simply look at the customer as a revenue entity in a seat, but as a human being with the potential of endangering the airline worker, other passengers and the safe operation of the aircraft. Although instances of passengers manifesting mental illness are rare, alcohol abuse and reaction to cumulative service failures are the predominant factors leading to verbal retaliation and abuse.

A fresh perspective on corporate responsibility in the changed social environment on-board is needed combined with the implementation of a much more passenger focused crew resource management approach in flight. Low-cost airlines do not have to equate with low-level passenger conduct on board. I believe that given the choice, the majority of air travelers would like to avoid embarrassment or at worst, costly fines, imprisonment, loss of job and family distress.

Cabin Crew Resource Prioritization

Figure 3.2 identifies criteria for cabin crews prioritizing resources within their own team and their passengers on a given flight. It focuses on prioritizing known special passenger needs and matching special cabin crew member skills with these needs.

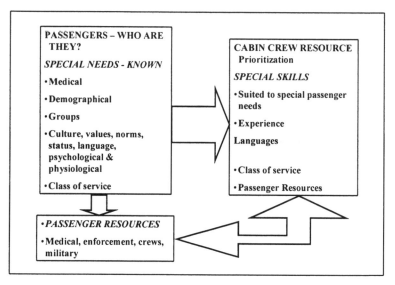

Figure 3.2   Cabin Crew Resource Prioritization

Three elements can assist the cabin crew in the planning and organizing stage of the preflight briefing for enhanced management of irregular passenger situations onboard:

- Identification of special passenger needs
- Matching cabin crew member skills and experience with passenger needs
- Identification of resources within the passenger group.

This model complements the On-Board Relationship model presented and discussed in Chapter 2. Equally important is the identification of passenger resources that could act as backup in the case of an onboard incident. The better the organization of such information is during the preflight briefing, the better is the cabin crew prepared to face the stresses of onboard conflict.

Identifying Special Passenger Needs

The following are some of the primary sources for identifying specific passenger needs prior to the cabin crew briefing:

- Marketing information (charter, route characteristics, groups, etc.)
- Flight manifest
- Corporate Security
- Bulletins
- Operations Control Centers
- Cabin crew experience.

Special needs identified in this manner are helpful in assessing what situations could occur. Crew resource management should be utilized in assigning cabin crew members with the best-suited skills and experience to the appropriate working positions. Overall response strategies, including backup should be discussed with the crew in preparation of the flight.

As part of their working routine, cabin crew members should implement the following activities:

- In the airport on route to the boarding gate, observe passenger behaviour;
- Be alert to unusual behaviour and notify the appropriate authorities;
- The In-charge cabin crew members should communicate with the gate agent prior to boarding the flight for additional passenger information;
- Brief the flight crew;
- Follow-up with the cabin crew;
- Always practice situational awareness.

A joint flight crew and cabin crew briefing is preferable, however is not necessarily operationally feasible due to the possible different contractual aspects affecting crew check-in times and flight boarding times. As a minimum, the In-charge cabin crew member must introduce him/herself to the pilot in command and review key aspects such as handling passenger misconduct should this occur, obtaining clarification of communication processes and expected support.

Giving preferential assignments based on seniority without matching cabin crew member skills and experience to the needs of a particular passenger flight profile, are not in the best interest of prevention. Promoting work assignments in this manner, does not always come easy. Company policies, contractual bid-positions, multi-national composition of many crews and established crew norms, are some of the aspects driving working assignments. Considering these constraints, cabin crew leadership is essential in preventing acts of passenger misconduct.

Cabin crew professionalism involves an intimate knowledge of passenger needs, a willingness to best fulfill them in the context of prescribed safety and service requirements, a conservative approach to the challenge of conflict, a sense of humour, and a time-effective response to problem solving. When everyone made their contribution according to their ability, no one can ask for more.

All this requires strong work ethics and self-discipline, a tall order for any organization and team leader to achieve. When practiced, as it is by many cabin crew members around the world, the outcome is a rewarding experience for all concerned.

Summary

Thoughtful treatment of the highly diversified air traveler is one of the top priorities for preventing social unrest on-board. The air traveler should not be left more aware of what is held back during the service experience than what is given. To be treated with respect and human kindness is part of the deal. The efforts to accomplish this are no longer the sole responsibility of the cabin crew however that of the airline organizations at large. Leaders willing to accept that formidable responsibility will integrate actions, measure and reward workers for their extra efforts.

It took time to develop crisis management for aircraft accidents, bringing all relevant departments together. Similarly, committing to a zero-tolerance policy must foremost be aimed at effectively preventing and manage passenger malcontent.

Notes

[1] *Calgary Herald,* 8 May 1999.
[2] *Calgary Herald,* 8 April 1999.
[3] *Calgary Herald,* 10 January 2000.
[4] *Calgary Herald,* 6 April 2000.
[5] *Calgary Herald,* 23 October 1999.
[6] *San Francisco Examiner,* 12 July 2000.
[7] *Airlinebiz,* 27 March 2000.
[8] *Airlinebiz,* 31 March 2000.
[9] *Calgary Herald,* 7 January 2000.
[10] *Calgary Herald,* 25 March 2000.
[11] *The Associated Press,* 6 July 2000.
[12] *The Associated Press,* 6 October 2000.
[13] *Calgary Herald,* 16 March 2000.
[14] *Calgary Herald,* 17 December 1999.

[15] *Avflash,* 7 February 2000, vol. 6, 06a.

[16] Investigating the psychological effects of the new height of the upper deck of the Airbus A3XX on human performance (passenger evacuation), factors such as visibility, slide design, passenger safety instructions, and social factors were also examined. Interestingly, subjects on the upper deck did not have a higher level of anxiety immediately before jumping than those of the main deck group, however the Exit Hesitation Time (EHT) was higher on the upper deck. A total of 115 subjects participated in this study, divided in groups of 16 to 20 for evacuation, with one cabin crew in attendance at the exit (Jungermann, Goehlert, 2000). As in other evacuation studies, women, especially over 50 years of age had a significantly higher EHT.

[17] *The Seattle Times,* 29 August 2000. Developed by Seattle based AquaJet Appliances in partnership with Boeing, the shower is the latest amenity available for consideration by airline marketers.

[18] *Financial Post,* 24 June 2000.

[19] Cabin crew behaviour in emergency situations has been the subject of research at Cranfield University. A program conducted as part of a memorandum of cooperation between the UK Civil Aviation Authority (CAA) and the Federal Aviation Administration (FAA) concluded that the behaviour and number of cabin crew 'significantly influences the speed at which volunteers are able to evacuate (Muir, Cobett, 1995).

[20] The FAA, responsible for the air traffic control system has not been able to keep up with the transformation of the industry since deregulation in 1978 and the economic growth of the last eight years. The FAA undertook plans to modernize air traffic control in 1981 with the Advanced Automation System as the centerpiece. In 1994, after the initial cost of US$12 billion had tripled, the results were disappointing; the project was still ten years from completion. The predictions of the National Civil Aviation Review Commission in 1996 have proven to be correct: '...rapidly growing demand combined with reduction in capacity, as a result of continued reliance on outdated equipment, will bring our nation's aviation system to gridlock soon after the turn of the century.'

[21] Critics point to the economics on which the management of air traffic control services is based in the U.S., and not pricing the services properly. The services are currently paid for by taxes on airline tickets, cargo and fuel. Users of commercial airlines pay the same as do private pilots, and small planes get the same priority as large airplanes. In Canada, air traffic control (NAVCAN) is privatized and completely self-supporting.

[22] Throughout aviation history, passenger comfort is relative. At the start, the only consideration was to keep the passengers alive (Emenaker Kovarik, Graeber and Mitchell, 1999).

[23] Human Factors considerations in aircraft cabin design are part of the aircraft configuration process (Emenaker Kovarik, Graeber, Mitchell, 1999) during the development of the overall aircraft design. At the early stage of design, engineering and marketing organizations aim for a 'configuration that is certifiable, economically viable, and competitive in current and future markets, with future derivative growth potential.' Design efforts involve trade-offs, including passenger comfort. The competitive environment expands the functions of the seat with 'human factors contributions in the areas of packaging, accessibility, multiuse interfaces, miniaturization, safety, and esthetics.' Brauer (1996) points out that long-range airplanes are designed with greater consideration for passenger comfort due to its importance to customers on long-haul flights.

[24] *Scotland on Sunday,* 4 April 2000.

[25] *Scotland on Sunday,* 4 April 2000.

[26] *Financial Post,* 8 April 2000.

[27] *The Globe and Mail,* 15 July 2000.

[28] *Avflash,* 10 July 2000, vol. 6, 28a.

[29] *Calgary Herald,* 5 March 2000.

[30] The physiological effects of oxygen deprivation combined with poor diet, lack of physical

fitness, stress and fatigue are some of the factors established by medical research (Vogel, 1995). Barometric differences between the cabin environment and inner body conditions can cause abdominal discomfort, diarrhea, and nausea. Crossing time zones, diet changes, and irregular working and sleeping hours are also reported to cause digestive disorders (Pieren, 1997).

[31] *Calgary Herald,* 10 October 2000.

[32] *Calgary Herald,* 28 October 2000.

[33] *Docket Number FAA-2000-7119-14,* 24 May 2000.

[34] *Docket Number FAA-2000-7119-1,* 24 May 2000.

[35] *Los Angeles Times,* 9 July 2000.

[36] ASHRAE sets voluntary standards for air systems.

[37] *HVAC/R Industry News – Press Release,* 27 January 1999.

[38] President of Consolidated Safety Services, Inc. in Fairfax, VA.

[39] Data was gathered during eight Boeing 777 commercial airline flights operated by an U.S. carrier. Four were domestic flights between 1,000 and 1,500 miles, and four were international flights greater than 3,000 miles. While airborne, cabin crews and passengers were asked to complete a comfort questionnaire that polled passengers on a variety of factors.

[40] *HVAC/R Industry News – Press Release,* 15 July 1999.

[41] *Air Travel,* 'Cabin Air Controversy', 30 June 1999.

[42] The full report is available on CUPE's website: airdiv-cupe.org/health.phtml

[43] *TSB Advisory #1388,* March 1990.

[44] *Smh News,* 22 April 2000.

[45] *Avflash,* vol. 6, 13b 2000.

[46] *Washington Post,* 9 July 2000.

4   Crew Members

Introduction

The air traveler recognizes flight and cabin crews belonging to a certain airline by the uniform they wear, their accessories and their grooming. They appear unified by their overall image while walking through the airport, mostly in small groups, a team in the eyes of the public. Air travelers expect 'their crew' to act accordingly and get them safely to their destination. Technical expertise from the flight crew is a given while efficiency and expertise to deliver the service and solve the human problems on board is expected of the cabin crew.

Flight and cabin crew coordination and communication is not an issue for the air traveler unless there is an irregular occurrence requiring optimal interaction between these two groups of professionals. This is however the case in aircraft incidents and accidents. Several research studies (Chute & Wiener, 1995; 1995b, 1996) and disruptive passenger incidents have identified certain deficiencies in this area.

The reasons for these deficiencies are rooted in historical, organizational, environmental, psychological, and regulatory factors (Chute & Wiener, 1994; 1995). Flight and cabin crew developed their own cultures, further inhibiting teamwork by the increasingly cultural diversity within both groups.

Although very specific recommendations have been made by NASA researchers to improve this situation, relatively little progress has been made overall. The focus for such improvements is procedural and training oriented such as in joint Crew Resource Management (CRM) training, while organizational issues are left to status quo.

In this chapter, I illustrate the different components and their impact on the relationship with the passenger with a closer look at the history of the cabin crew profession. Cabin crew communication and coordination receive special attention.

Technical advancements in aircraft design are closely linked to the changing roles of both the flight and cabin crew, including problems with communication and coordination between these two groups.[1]

A Historical Perspective

When civil aviation emerged after World War I, transforming cargo operations to expand into passenger transport, the German airline Lufthansa recognized the commercial benefits of customer service. Albert Hofe was hired in 1928 to become the first airline steward in the world serving Lufthansa's passengers on its prestigious Berlin to Paris route. The role of the steward, modeled after the role of a ship and train steward, was strictly dedicated to attending to passengers' personal needs. An airline steward was authorized to order whatever supplies necessary to satisfy his customers from the privileged ranks of society. Well versed in etiquette, creative in problem solving and impeccably groomed, he embodied the new professional adventurer in a much sought after new career.

Safety training was not an issue in the early days of civil aviation. The steward was expected to follow captain's orders in case of an emergency.

Women were not granted access to the service side of the budding air passenger transport industry in the United States until 1930. It was a matter of chance encounter between Steve Stimpson, District Manager of Boeing Air Transport (BAT) and Ellen Church, a registered nurse with ambitions to become a pilot. Church visited his offices in February 1930 resulting in Stimpson making an innovative proposal to employing women instead of stewards. The idea of introducing women into an exclusively male work environment was not met with immediate enthusiasm. Church became the world's first stewardess on May 15, 1930 on a cross-country flight from Oakland, California to Chicago. After a trial period with eight trained stewardesses on the Chicago-San Francisco route, a new career was open to women, fondly called 'sky girls'.

From Sky Girls to Cabin Crew

The nurse's profession provided the airlines with its first candidates for the new career, the airline stewardess. Nurses became the Nightingales of the skies. They were ideally suited for the airlines' needs at that time. Trained in handling emergencies, they were comforting towards people who were afraid or did not feel too well, competent to deal with unusual situations, and dedicated to the welfare of anyone under their care.

On September 1, 1937, Trans-Canada Air Lines (TCA) operated its first passenger and first international flight from Vancouver's Sea Island Airport to Seattle. The aircraft was a 10-seat Lockheed 10A Electra without a jump seat for cabin crew. Only nine passengers were carried with one seat reserved for the stewardess. The cabin was small with five seats on either side of a narrow aisle, and not suited for anyone over five foot and five inches to stand up. Because of the low altitude flying in a non-pressurized aircraft, ear problems and turbulence were common occurrences, and no passenger would dare to get-up. There was no provision for an on-board galley. Coffee was served from a large thermos bottle; newspapers and magazines along with complimentary cigarettes and chewing gum were the standard fare.

The role of a stewardess was to alleviate passenger concerns and fear of flying with explanations of aerodynamics, cloud formations, and meteorology.[2] They also acted as tour guides since large windows offered spectacular views of the landscape below at low altitude flying of 10,000 feet or less.

Stewardesses looked after the entire passenger process from boarding to deplaning. They would meet the passengers in the airport and personally introduce themselves to everyone. Eva Mossop, a veteran with Air Canada and now retired, recalls in a discussion with the author:

We got to know our passengers by name, including personal details about their families and work since most of them turned out to be regular customers. We went into the boarding area and met with our passengers. We would observe our passengers and knew right away who would be difficult. We would talk to them and find out if we could help. If the person appeared in rough physical condition or inebriated, we would explain that flying would make things worse and mention it to the pilots and the passenger agent. Cases were rare when we decided to leave someone behind, and we handled them discreetly without upsetting anyone.

Because of the restricted space, carry-on baggage was not allowed. Early seat design did not provide any room either for under seat stowage. The overhead netting served for pillow, coat and hat stowage only.

Liquor and drunk passengers were not a problem. There was no liquor on board. The few passengers who occasionally brought out their own flasks were quickly and politely discouraged. Rules were rules, and the nursing background of the stewardesses prepared them well to deal with the rare case of a budding conflict.

The novelty of flight, the small group of people utterly dependent on the stewardesses for reassurance and physical assistance, created a very personal relationship between the crew and its passengers. The needs were basic, matched perfectly with the skills of the Nightingale of the skies. The stewardesses were respected as competent women in good part because of their professional background as nurses. They were respected for being part of a daring new workplace, previously the exclusive domain of pilots and flight engineers; it was the most glamorous and envied career a women could have at that time.

Selection criteria for stewardesses still focused predominantly on appearance and social skills in the late 1960's (see *Appendix E)*. Since one of the hiring requirements stipulated that stewardesses had to quit once married, the turnover was high. The average tenure of a stewardess was approximately two years, not enough time to get into a routine or become disillusioned and callous. This is in stark contrast to cabin crews in North America today, where Human Rights legislation has lifted the many restrictions of the early times of aviation, and employees are staying on in great numbers until legal retirement age.

Other restrictions such as weight limitations were cause for disciplinary action, forced weight programs and in the most persistent case, cause for dismissal. Battles of the past continue even today as evidenced in the Frank vs. United Airlines case. The issue is the legality under federal and state law of the 'flight attendant weight program' in effect between 1980 and 1994.[3] The program was implemented with the approval of the union at the time and established two different weight requirements for males and females. An earlier ruling rendered in 1979 on the same issue, concluded that the airline's program guidelines were lawful. In June 2000, a federal appeals court ruled in favour of the female cabin crew members who recently sued the airline over the discontinued program. United Airlines is appealing this ruling.[4] Clearly, the challenges for the cabin crew members, their associations, and management are much more complex today. The relationship between these parties often plays an important part in the air travelers' service experience.

Changing Relationships

Basic physical safety needs of the early air traveler fostered a strong bond between cabin crew members and their passengers. Social norms of civility and social conduct were a given. The emphasis on professional attire and communication was reflected in the airlines' selection criteria, training programs and its performance standards. The mature professional women with nursing degrees, a highly developed sense of social etiquette and conduct within society including the acceptance of authority led to a well functioning social environment on-board. Cabin crew members were exclusively assigned to passenger service duties for a couple of decades before their safety role in emergency situations was recognized. Veteran cabin crew members reminisce with fondness when they passed time with their passengers during delays, playing bridge or cribbage, taking them to movies, lunch or sightseeing.

The introduction of larger aircraft resulted in shifting on-board relationships between the cabin crew and the air traveler. Without any preparation to speak of, the on-board dynamics changed. Organizing cabin crews into formal teams with defined roles and responsibilities changed their autonomy previously enjoyed. Working life became a bit more complex affecting the social dynamics between cabin crew members and their interaction with passengers. Whatever happened on-board from now on, happened in a changed social environment: cabin crew and passengers alike were now constantly maneuvering through a cast of strangers to have their different needs and expectations met.

The Need for Partnerships Beyond the Passenger Cabin

With the rapid growth of the aviation industry after World War II, cabin crews started to organize themselves in order to have some input into their working conditions, wages and benefits. A new layer of communication was added between the cabin crew and airline management; from now on, union officers represented stewards and stewardesses in these matters, and personal negotiating with one's supervisor or manager seized to be an acceptable practice.

Since then, the relationship between management and cabin crews has not always been smooth. Together with flight crews, they enjoy an arms length relationship with management, physically distant compared to their colleagues in other departments, autonomous, and little supervised. Although cabin crew member labour associations are only infrequently resorting to strike action, even the threat of a strike has major financial consequences for the airline, and is deeply upsetting to the traveling public.

Figure 4.1 illustrates the growing layering of the relationship filters between the cabin crew and the air traveler. The past is illustrated on the left hand side, the present on the right hand side.

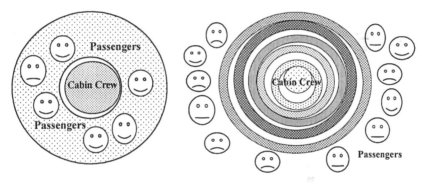

Figure 4.1   **Cabin Crew-Passenger Relationship Filters**

Working conditions and benefits became increasingly an issue for cabin crews as the industry grew in the 1940s. Close bonds with flight crews, management and passengers were no longer the most important issue as Muriel Peacock, founder and first president of the Canadian Airline Flight Attendant Associations realized. Growth changed relationships from the personal to the impersonal. Flexibility in negotiating individually for days off, vacation and sick leaves with a supervisor became unmanageable. The need for an official bargaining unit became clear.

Other partnerships evolved with regulatory authorities as a result of cabin crew labour leaders using the umbrella of the International Labour Organization (ILO). These efforts culminated in a submission to the International Civil Aviation Organization (ICAO) in the early 1960s to recognize 'flight attendant' or 'cabin attendants', as essential members of the crew. As regulatory authorities responded to a changing industry, air travelers saw their on board conduct increasingly restricted. Non-compliance with safety regulations became one of the recognized causes for on-board conflict.

The United States led the way in 1952 through the Civil Air Regulations, establishing the 1 to 10-passenger ratio. The introduction of larger aircrafts resulted in an amendment to these regulations in 1965 to provide one cabin crew member for every 44 seats. In Canada, the Department of Transport followed suit by drafting the Air Navigation Order, Series VII, No. 2, when the category of stewards and stewardesses was recognized in their safety role under the new designation of flight attendants or cabin attendants.

Change does not come about without driving external forces. Drastic events are more likely to shake complacent attitudes. A string of fatal aircraft accidents both in the United States and Canada over the period of 1965 and 1966 combined with the positive role of cabin crew labour associations in pressing for regulations cannot be underestimated.

Although regulatory recognition of the safety role was a major step forward, airlines' selection criteria did not reflect skill requirements that might have supported this role. The emphasis persistently stressed appearance and congeniality of the female applicants, as the reader will see in an example of an interview assessment form used in the late 1960's (*Appendix E*). Uniform styles in the late 1960's and early 1970's reflect the airlines' denial of the emerging safety role of cabin crew members and the predominant value of placed on marketable image.

The efforts by union leaders to raise safety awareness in their own ranks was fraught with difficulties, not the least because of the existing value system and role perceptions of professional nurses who strongly objected to unionization. N. Jill Newby, the author of the history of the Canadian Airlines Flight Attendant Association (CALFAA) documented this struggle in her book 'The Sky's the Limit'.

Crew Complement and Service

In Canada, the minimum cabin crew complement is regulated based on the emergency evacuation model with a 1 to 40 cabin crew member to passenger ratio. Differences exist in different countries. In the U.S. for example, the ratio is 1 to 50 passenger seats fitted on the aircraft. The cabin crew complement for service considerations is each airline's decision based on a range of criteria, not the least cost and competition, cabin configuration, galley placement, flying time and classes of service.

In the competitive airline industry labour cost receives the same scrutiny as unit cost for meals and all other service amenities. For airlines operating in the highly industrialized North American and Western European countries, labour cost are significantly higher than compared to other airlines in other regions. Individual contractual requirements add or diminish management's ability for flexible manning of their flights in an effort to control cost. The issue of what is considered an adequate cabin crew complement has and continues to be a recurrent issue in labour negotiations. The higher the ratio of passengers to cabin crew, the more complex the service activities are to be performed along a given timeline, the less time for personal relationships with the air traveler is available.

Examples of cabin crew to passenger ratio are illustrated in *Figure 4.2*.

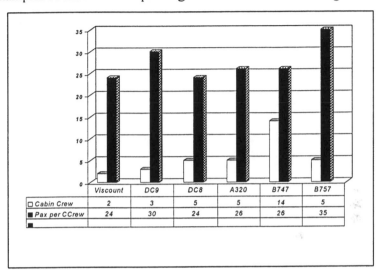

	Viscount	DC9	DC8	A320	B747	B757
☐ Cabin Crew	2	3	5	5	14	5
■ Pax per C Crew	24	30	24	26	26	35

Figure 4.2   Passenger-Cabin Crew Ratio Averages

Although crew complement may have some influence on the time available to deal with individual passenger conflict situations, it certainly is not a ready argument for increasing the number of cabin crew members. There is currently no study that would offer any evidence of a relationship between the number and types of disruptive incidents on board and the total cabin crew complement.

Service Expectations

Meeting, Exceeding, Anticipating Customer Needs

The service provided to passengers has undergone spectacular changes over time affecting every aspect from cabin design and configuration to amenities and food and beverage service and finally to the role of cabin crew members. Driving forces were new technology and the evolving customer base from one of the privileged few to mass travel for all levels of society. Airline service has had to increasingly reflect the diversity of the global customer, unless the route network is restricted to the needs of a more regional cliental. Marketing strategies such as incremental innovation are part of the fast-paced global competition in response to the constant change of mini-market needs in the passenger cabin.

Ken Day, a 40-year veteran of the Canadian aviation industry reflects on service attitudes in an interview with me:

> If people did not have certain qualities when they were hired, they were not likely to develop these skills later during their career. Some, even in the early 1950s expressed their views in statements such as this: 'If a passenger is rude, that justifies my being rude.' There was a difference in service attitude when comparing North American hires compared to those from Europe. North Americans were not as tuned into a server's role than their European colleagues.

> Yes, we had the isolated incident of passenger misconduct, usually prompted by overly generous cabin personnel dispensing alcohol too liberally.

Eva Mossop who joined Trans-Canada Airlines, now Air Canada, comments on flying in 1986:

> Flying now is easier in some ways than when I first started (May 1, 1939), but the contact with the passengers is much more distant – we haven't got the time. But passengers have changed too. They used to ask us where we live, and about our life; now it's very impersonal. A lot of the glamour has gone. We say it's like a bus service now: we had a lady the other day with curlers in her hair! ... It has really changed remarkably...

In the middle 1980s, the marketing slogans advocated that customer satisfaction could be achieved by meeting customer needs. Exceeding customer expectations was the next promise, followed by anticipating customer needs. Creating needs has become a new driving force in product development and marketing. With the number and variety of tangible products available onboard, so does passenger malcontent when these services are not available. One of the newest service enhancements being contemplated for some time is the introduction of on-board video gambling. .

Time is of Essence

Time parameters drive the in flight product delivery. Limited cabin crew resources versus passenger load can lead to conflict with safety related regulatory requirements.

Even with all efforts to provide the cabin crew with the tools to do the job, time is of essence. On a flight of fifty-five minutes ramp to ramp, the actual time available for delivery of a breakfast for example is twenty-five minutes in between ordinance signs. With a passenger load of one hundred and twenty passengers and a cabin crew of four that leaves less than one minute per passenger and cabin crew interaction. Even the most basic request a passenger makes infringes on the precious seconds the cabin crew has to deliver the standard product. The cabin crew will experience these extra requests as stressors and in potential conflict with their primary tasks.

Failing to reconcile the safety and service role of the cabin crew on short-haul, highly competitive routes may happen. Service products must meet the test of the operation. In case of doubt, the operational mind should win the day.

Pressed for Problem Solving

When time is of essence, a passenger in need of problem solving soliciting intervention by the cabin crew will jeopardize the service delivery to the majority of the passengers, thereby creating another problem. There is hardly any other work situation where product delivery is fraught with such risk taking. In any other production environment task bunching of this kind would be unacceptable, and the planners would be held accountable. Unfortunately, this type of management action is not uncommon in the airline industry. It is symptomatic of a style that does not take into account the full range of traveler needs begging to be satisfied. It defeats all claim to quality service, and can place cabin crew members in a robotic state.

I have found in my research that these conditions result in a consistently higher number of complaints, with a lack of empathy towards a passenger's plight taking the lead. Passengers are generally sensitive to cabin crews' predicaments on these types of rushed, service dense flights, however passengers should not be placed in a position where they feel sorry for the cabin crew.

The Problem of Too Much Time

The other extreme occurs when overly generous cabin crew staffing results in lengthy periods of unproductive time, and service delivery is not an issue. Long-haul night flights are one of the examples. After the initial service rush, most of the passengers tend to prefer to sleep, and the cabin crew is under-utilized.

Exceptional cabin crew teams excel in being vigilant and keep a discreet cabin presence throughout the service downtimes. They are acting in the full recognition that a satisfied passenger is the key to preventing problems. They are miraculously at your side when you open your eyes and feel parched. They adjust your blanket and pick up any reading material or debris from the floor. They monitor mothers with children, the elderly and the restless ones, finding ways to assist before they are asked. This constant but measured pacing of activities also counteracts fatigue, since night flights interfere with their circadian cycle.

On the other hand, their less motivated colleagues cater to their own needs first. They gather in the galleys and chat, usually loud enough to be overheard in the cabin by the sleepless and weary passenger. Innocent as this practice may be, publicizing details regarding personal events, schedules, job dissatisfaction, relationship and personal crisis do not contribute to customer satisfaction.

The question of filling service downtimes with meaningful activities is not that simple. Today's air traveler enjoys a multitude of choices to occupy time: by watching movies, listening to audio entertainment, using laptops for work or pleasure, reading or sleeping are some of the many alternatives. These choices disconnect them from their environment, the other people and further impoverish opportunities to interact with cabin crew members on a broader social basis. Senior cabin crew members deplore of that loss and reminisce on the times when they were able to connect with their passengers on a more personal level.

Rote versus Engaging Performance

When there is service downtime, passengers have different expectations. If they are not occupied with reading, watching a movie, playing a video game or working on their laptop, all their senses are focused on the cabin crew. They watch what they do, time their absences in the cabin, and continuously formulate positive or negative perceptions.

The point is that cabin crew members, just as everybody else, need compelling personal reasons to turn the delivery of rote tasks into engaging performances. The exceptional performers do not just go through the motions. They continuously use all their senses and thought processes in assessing the subtleties of rapid and at times unpredictable changes in passengers' moods. They master and reconcile such apparent opposites as safety and service. They often act intuitively and delight in finding creative solutions to unusual problems. They maintain a fluid and flexible mindset and cooperate with one another within the well-defined framework of a flight. The impact of such spontaneously choreographed service on social peace in the passenger cabin is tangible. The concepts of leadership and teamwork in the passenger cabin will be addressed later in this chapter.

Jan Carlzon, past CEO of Scandinavian Airlines, popularized the expression 'moment of truth'. The 'moment of truth' applies to the encounter between the customer and the representative of the airline. In the service industry, acting with empathy is imperative, especially when things go wrong. Granted, facing up to a customer to advise him or her that there is no meal or no more choice is not an easy task. Cabin crew members do not enjoy apologizing for shortages they have no control over. Established airlines have policies in place to deal with such operational problems. How they are implemented is another issue.

A cabin crew team that is empathetic will do all possible to minimize the degree of dissatisfaction. Looking for alternatives, they might be able to put something together from first class they may give up their own crew meals, always with a word to the customer that they are working on a solution. Identifying standby and contingent passengers to avoid inconveniencing preferred and full fare customers, is one way to deal with known meal shortages.

I have come across a wide variety of cabin crew member responses ranging from not serving anyone at all to giving up their own meals to purchasing something during the next station stop and/or alert the next station of the problem, and be ready to compensate the passenger on arrival.

The exceptional cabin crew member performs independent of how well the organization treats its workers. It is a very personal attitude stemming from a genuine altruism and a touch of self-preservation. An organization recognizing its employees' deeds of empathy and caring will benefit in the long run.

Antagonizing Customers

The tragic death of a 19-year-old man, a passenger on flight from Las Vegas to Salt Lake City August 11, 2000, raised new questions concerning the role of cabin crew members faced with a severe case of passenger misconduct.

According to reports,[5] the young man became combative shortly before landing, screamed obscenities, hit several passengers, and broke a hole into the cockpit door before being subdued by passengers. The autopsy report concluded the young man died as a result of injuries inflicted by passengers restraining him.

Passengers witnessing the event have since reported their observations to the media. Comments range from criticizing the cabin crew for a public address announcement antagonizing the young man to failing to intervene when a passenger continued to leap repeatedly on the apparently unconscious man's chest.[6]

Extreme as this case may be, It reveals the urgency of devising appropriate procedures and crew training with the input of experts from various disciplines.

Change and Employee Motivation

Organizational changes in crew meal policies and quality can result in attitudinal changes, e.g. to shortage situations on their flights. When Air Canada was in a cost cutting phase, all aspects of expenditures were examined, including crew meal policies, cost and meal allowances. After investigating claims for meal allowances based on meal shortage reports a small minority of cabin crew appeared to abuse this situation. Certain individuals would submit claims that could not be reconciled with quality assurance reports, catering and load factor data. The resulting changes did not achieve satisfaction for a variety of reasons. To have a meal or snack boarded where previously a meal allowance was the norm, directly impacted on what cabin crews consider incidental revenue and manifested itself in the way meal shortages were handled.

In-charge cabin crew members who previously could count on their team members' ready cooperation now established in their pre-flight briefings who would be willing to give up their meals should there be a meal shortage. The level of volunteering declined, especially since claims previously approved would be rejected on a more frequent basis. On flights where meals or snacks were boarded and the meal allowance had been cut, cabin crews were forced to make tough choices, more often than not to the detriment of passenger satisfaction. On one hand costs were cut while on the other passenger complaints and the cost of compensation to passengers increased. In summary, management decisions and attitudes towards issues cabin crews care about have a significant impact on how passengers are treated. Research into the best performing service industry (Berry, 1995) in the U.S. confirms the benefits of this approach.

One classic account by Charles Garfield of his experience on a flight on Eastern Airlines before 'Air Rage' became a topic tells of a cabin crew member advising him that no meal was available for him while he was traveling in First Class. While his seat companion, sharing his fate, ranted and raved for some time, Garfield decided to stretch his legs and discovered two cabin crew members eating in the galley what looked like the First Class entrees. He recounts the reaction: 'Well, we've got to eat too, you know. 'Yes', he shot back, 'but you could have shared with us.' He was obviously enraged, especially since he was prepared to accept that the shortage was probably not the cabin crew's fault. The response however left him completely disillusioned with the airline. He also does not hesitate to link the cabin crew members' uncaring attitude to that of their corporate leader at that time, Frank Lorenzo.

While incidents of collective cabin crew compassion towards passengers in extraordinary situations such as aircraft incidents, illness and distress is rarely an issue, empathy is much less practiced in situations cabin crew feel are a regular part of service break down. This is evident from typical comments such as: 'They should know better. It happens all the time.' Responses such as these reflect on the desensitization cabin crew members either have developed due to job routine, the organization's lack of focus on service quality, or at worst represent their own personal feelings towards other people's measure of distress. No matter what the reasons, showing empathy towards an aggrieved customer is the smart thing to do, for personal reasons, not the company, not the passenger. Empathizing protects cabin crew members from potential escalating hostility and leads to greater success.

Dealing with Technology, Dealing with Emotions

Although the aircraft is the commonly shared workplace for both flight and cabin crew, there are fundamental differences between the physical environment they work in and the culture of these two groups (Chute & Wiener; 1994; 1995). The flight crew operates in a highly technical environment, isolated from the rest of the aircraft cabin(s) by the cockpit door, the cabin crew and the passengers.

Depending on the airline organization, there could be gender, age and demographic differences to start. Flight crews are still predominantly male, while cabin crews are predominantly female. Flight crews are highly technically competent individuals, an aspect that is firmly entrenched in training and operational disciplines.

By nature of their different job functions, flight crew enter commercial aviation after already having made substantial commitments by personally investing in costly training to gain their wings. This is not the case with cabin crews.

Cabin crews on the other hand perform mostly service function on-board, at least in the eyes of the passengers. Social and language skills are key together with a pleasing appearance and manners to enter this profession.

The cabin crew interacts mostly with each other in the service to passengers. They are highly mobile throughout the flight although restricted by the configuration of the cabin to a predominantly forward and aft movement. The passengers are facing forward just as in a theatre without a center stage waiting for the performance to begin. The aircraft stage is nothing but a narrow aisle cutting through the rows of seats front to back, usually offering only partial viewing of the performers, and often only of their backs as cabin crews offer service from their trolleys or trays. The picture of cabin crew members in aprons and with garbage bags in their hands for cabin clean-up is not conducive to the professional safety role associations wish their members to be identified with.

Galleys and washroom areas are confined traffic areas, defended by cabin crew and infringed upon by restless passengers. They are spots to mingle with the crew and other passengers. They can be places to unwind and places of increased tensions.

Motivation and Career Orientation

While flight crews have already committed to a profession with a focus on technical expertise, cabin crew members at large enter their profession on a very different note. The hundreds I have come in contact with and asked the reasons for their choice confirmed three basic reasons. Most of them confirmed that they joined the airline with the intention to do the job for a number of years, as a change from the drudgery of previous employment, or to gain clarity of what to do with their university education or simply because of the attractive life style the airlines afford them. A common thread of non-commitment to applying their skill to the demands of a profession they have been previously trained for is evident, which explains the difficulties both management and associations have to solicit candidates for positions in their ranks. Being in management requires commitment to the daily stresses of working in an industry that has and continues to be under great pressures to perform better. At the end of the day, these professionals take their work home, their minds enslaved by ongoing work issues. Cabin crews on the other hand, after completing their flights, take their uniform to the cleaners, and their minds are their own.

Once the life style and accompanying benefits have seduced individuals unsuited to dealing appropriately with the demands of the irregular work schedule and the diversified air traveler, the predominantly unsupervised environment offers little opportunity to dismiss poor performers.

Leadership Role – Cabin Crew

Since the basic cabin crew organization has its tradition in the structure of military and marine authority and command models, much of this is carried through today in cabin crew classifications and on-board assignments. Classifications enshrined in contracts can result in regular cabin crew to abdicate all decision making, using the convenient excuse that 'they get paid the bucks, and I am not.' This type of comment indicates a complete lack of motivation and is a clear measure of an unhealthy work culture. Coupled with abandoning all ownership to be up-to-date on company policies and procedures, product knowledge and service procedures, team performance and employee moral suffers.

In the example of one major Canadian carrier, the In-charge or 'Purser' classification was introduced in 1971. Originally, it was male dominated since the leadership position was gender based. Also in that year, the Flight Service Director position was created to address the introduction of the B747 into the carrier's fleet with the Purser in a secondary In-charge role.

Cabin crew members' opinions regarding the value of this position and its impact on the quality of their work lives varied greatly. Some resented the loss of perceived ability to perform public relations with the customers that would conflict with the role of the Flight Service Director. Others resented the lack of service involvement and the perceived additional workload as a result of instituting this new position. Still others refused to solve passenger complaints and left it up to the Flight Service Director to deal with a dissatisfied customer.

The position of a Flight Service Director was a much touted and prestigious classification without any service involvement other than public relations, flight-related administration, problem solving responsibilities, and duty free sales. Over time the role of the In-charge eroded from an initially highly respected, and clearly defined one to a more diluted role with little management responsibility and increasing service involvement.

Customers also perceived this gradual change and placed less value on the impact Flight Service Directors had on the quality of service from their viewpoint. Management's initial expectations relative to the people management did not materialize. Since the Flight Service Director classification was part of CALFAA, the conflict of being part of the same union as their co-workers, and being accountable for the performance of their teams, complicated matters.

Developments within the union added stress on cabin crew relations through the forced integration of male and female In-charges in 1973. This move left deep divisions within the classification as a whole, some of which were played out on board for years to come. Much depended on the individual quality of Flight Service Director leadership style and their relationship with their crew.

The two In-charge system on wide-body aircraft with at times blurred lines of functional responsibilities caused conflict and deep divisions within work teams as well as some tangible benefits for the organization. Over time it became more difficult to effectively measure the performance benefits of this classification such as evidenced in onboard complaint handling, boutique sales objectives and general team coordination.

Subsequent to the cabin crew strike in 1985, the thrust of the negotiations targeted a reduction of the Flight Service Director role. Management's lack of support to the Flight service Director group was partially responsible for the failure of this classification. Lack of management resources and difficulties to manage the qualification program were some of the main deficiencies. Ultimately, the Flight Service Director classification was phased out, leaving the current system of a single In-charge on all types of aircrafts. This move reduced costs and improved scheduling flexibility

Authority and accountability issues continue to be at the center of cabin crew work teams. Some airlines have a management representative with overall responsibility for the performance of the cabin crew. Without the opportunity to examine the impact on team performance and passenger satisfaction indicators, I cannot comment on the effectiveness of this model. Cultural issues and history of the organization will always have an impact on the validity of such structures in the current market environment. Business today, especially in the industrialized world, is increasingly retreating from the model of the pyramid structure and is evolving into a more democratic, team based organization. The trend is to give the worker more control to make decisions affecting their work while at the same time providing access to all relevant information and basic guidelines for decision-making.

Teamwork

Teamwork is a much-heralded concept and credited for the success of many companies, especially in the manufacturing sector. Entire organizations have been structured along team principles, with the aim to foster employee participation, and partnership and in the process achieve greater productivity and quality output. Teamwork is a central theme for flight and cabin crews however aircraft accidents and incident reports identify on-going problems. These problems are not restricted to the interface between the flight crew and the cabin crew they also manifest themselves within the cabin crew itself.

As Charles Garfield describes: ... 'the team ... is not a fixed work unit, not a rigid entity, but rather a fluid, flexible *pattern of movement* of people and resources over time.

Teamwork is essential in the new story world in which speed, creativity, and flexibility are prerequisites for success and constant change is the context in which the organization operates.'

Although the label *team* is liberally applied to cabin crew work groups, it does not mean that they automatically function as a team. Just the same as the organization they are part of, they may reflect a top-down, autocratic leadership, with fixed relationships and work positions, and where pettiness stifles flexibility and creative problem solving.

The demographics of cabin crew teams can be highly diversified; ethnic and cultural mix, different mother tongues, social, cultural and religious backgrounds, mixed with diverse levels of education and work experiences make for a highly heterogeneous group. This fact alone places a high demand on leadership skills, necessary to forge the multi-talented individuals into a team.

Airline mergers have added further stress on team relations, when seniority lists are merged, and cabin crews coming from different organizational cultures suddenly have to work together. Admittedly, this can lead to deep-seated animosity and crew conflict.

Cabin Crew Team Challenges

Flight and cabin crews belong to separate unions. Contractual issues affecting aircraft qualification, scheduling, duty times, or hotel accommodation are subsequently different for these two groups (Chute & Wiener, 1995b). While there is a clear career path for flight crew to develop their technical and leadership skills on smaller to larger equipment, from right seat to left seat, the same does not apply to the cabin crew.

Scheduling needs, labour contracts and bid preferences governed by seniority, including marketing requirements for language qualifications, class of service or galley specialization, are some of the factors driving the team composition. Airlines require flexibility of staffing in order to deal with operational irregularities, weather delays, fluctuating load factors and cabin crew member absenteeism. A reserve pool of cabin crew members, combined with seasonal staffing is a common strategy the airline industry employs to maintain adequate coverage of flight requirements.

These methods, designed to optimize staffing and operating cost ratios have drastically changed the nature of the on-board team since the early days of civil aviation. The cockpit crew and the cabin crew stayed together for the duration of a scheduled period, creating the opportunity to form strong bonds and team loyalties (Chute & Wiener, 1995a).

Cabin crew member qualification on multiple aircraft types opened up a new type of team concept. Although more cost efficient due to the flexibility of scheduling to assign cabin crew members where needed, the traditional team bonding with the flight crew disappeared. In the example of Air Canada, there is no restriction on a given aircraft type. Cabin crews are qualified on all aircraft types within the fleet. Block holders may work different aircraft types during a given block month while a reserve cabin crew can be called upon to work different aircraft types and with different block holder teams during any given workday. This type of work environment calls for a great measure of flexibility, not only by the reserve cabin crew but also by the In-charge cabin crew members.

The In-charge crew member must be able to integrate the newcomer effectively into the existing team on short notice. The team members have often no prior knowledge of the skills and capabilities of the new addition to the team. The briefing is crucial to assist in the integration process, however it is often restricted to just minutes, barely covering some cursory reference to emergency procedures and on-board service aspects. Planning and organizing skills together with strong communication skills are prerequisites to facilitate the integration of new cabin crew members. Complete reliance on the In-charge for detailed product and service reviews no longer has a place in this fast moving environment.

In unionized organizations, the entry requirements for the In-charge position are not subject to experience and demonstrated leadership skills and other performance criteria such as problem solving skills and conflict resolution. It is a bid position, awarded by seniority allowing the most junior cabin crew member to enter the classification after qualifying in a specialized training course. There is no common standard that guides this training in the industry.

The position is no longer viewed as desirable in many airlines since the pay differential is minimal while the nature of the work tends to be viewed as unrewarding, fraught with 'hassles', additional administrative duties, and generally increases the potential for performance related follow-up with management (Chute & Wiener, 1995b).

A more senior team with a junior person in-charge frequently experiences difficulties in resolving leadership issues on-board, resulting in a lack of crew coordination and communication with a negative impact on customer service and conflict resolution.

The Exceptional Team

A cabin crew team can achieve extraordinary miracles under pressure. Left with no resources other than a genuine caring for each other and their customers, their creativity and intuition combined with a will to survive under deteriorating conditions on board, can turn a hopeless situation into an outstanding success.

The most exceptional teams function like an organism, they excel at self-managing, self-renewing, and self-transcending. For cabin crew teams, that translates itself into every member being knowledgeable and up-to-date on all aspects of their work. It requires autonomy of the team and the individual team member who organize themselves along the needs of a particular flight profile, and who make decisions, solve problems quickly and efficiently based on a common value system, and good communication skills.

Most cabin crew members in the industrialized countries operate at arms length from management. Rarely does management come on board to specifically supervise a flight. This lack of control creates a level of anxiety in management ranks, especially in the relationship between Inflight Services and Marketing, with Marketing wanting to see their products delivered in a consistent manner.

The Challenge of Performance Measurement

With the reduction of management levels by air carriers, new forms of performance measurements were introduced, mainly voluntary completion of observation forms by colleagues traveling on business, and customer surveys, specifically designed to measure crew performance in the areas of product delivery and customer relations. Ghost riders, (individuals employed occasionally by an airline to gather crew performance data) are also adding to the performance data pool. No matter what the source, the problem of acting upon unfavourable ratings is one of timing and management manpower. The disciplinary process is lengthy with very few cases resolved in the interest of the company or the customer.

Despite schedule fluctuations from one bid period to another, and seniority based classifications, exceptional team members tend not be possessive about their functions, positions, or work assignments in the passenger cabin. Since work rotation is part of these team's norm, there is little opportunity for rivalry and up-manship.

Self-managing also includes regulating members with performance problems. This is particularly difficult in some unionized environments where any form of self-regulation is considered a part of discipline and therefore management's responsibility. The consequences of being isolated from the support of either union or management is one of the deterrents in self-regulating when the overriding culture is opposing such actions.

Self-renewal is meant in the sense that in a well functioning organization, the value of teamwork is commonly appreciated to the degree that newly formed teams immediately adjust to the task at hand. Members are flexible and competent and organize themselves quickly along their special skills to deal with both operational and passenger relationship issues. On closer examination one discovers members of highly functioning teams having developed strong bonds within their community through such activities as volunteering, involvement in business, politics, or by furthering their education.

Self-transcending describes a team able to create its own legend. This team thinks together, gathers input from customers, evaluates the information and passes it on to the next level for consideration. This team exchanges ideas and celebrates its successes as much as it learns from its failures. Its members are respected by their colleagues at all levels of the organization, are welcome when they link up with other teams and often take on the role of a coach.

The concept of self-managing teams is not something that can be readily adopted in every culture or in every organizational setting. In non-western cultures where strong social and cultural norms guide structure and organizations, service orientations and service quality are based on general values associated with service functions and society's value of the service provider. Forcing organizational changes to reflect western models and values when no support exists in the larger environment of the home culture is fraught with enormous obstacles and conflict.

A New Model of Emergency Response

New emergency procedures at Air Canada after the Cincinnati accident in 1983 are another example of changing from a hierarchical structure to a self-managed team. Transport Canada initiated a Working Group to develop a new concept based on accident response to occurrences with no or little warning. The changes to procedures was dramatic since the new concept involved decision making by each cabin crew member to assess the situation according to a time line rather than on the traditional approach of assuming there was time to prepare. After decades of following procedures learned by rote, cabin crew members were now expected to think situations through according to a set of priorities.

The impact of this change was visible in a reduction of the volume of emergency procedure manual in a number of ways, including the elimination of aircraft specific emergency evacuation procedures. Previously there was a requirement to issue a new card each time the aircraft configuration was changed. On average production time for these cards was six week before they were ready for distribution and procedural implementation. The outcome of these changes resulted in two generic cards, one for the number one emergency position and one for all other cabin crew positions, subject only to incremental changes since then. Naturally procedural changes had to go hand in hand with the standardization of emergency equipment on Air Canada's fleet.

Resistance to the changes had to be managed with a carefully developed communication and training strategy, since with this change, the comfort of old assumptions was suddenly eliminated. A shift in roles and responsibilities had to be addressed, not only within the cabin crew ranks but also in the relationship with the cockpit crew. Greater autonomy of the individual cabin crew member could only work if communication processes were clearly defined to support this change.

The effects on the quality of cabin crew accident response capability have been no doubt contributed to the advancement of cabin safety. A clearly defined shared mission, empowerment, flexibility, and equal partnership, are at the center of these changes.

Flight – Cabin Crew Relations

The effects of changing social attitudes towards authority figures is also felt in the relationship between flight crews and cabin crews, especially in the industrialized countries. Adding to the growing alienation are the divergent schedules for both parties and the trend of maximizing schedule efficiency, allowing little time for a joint briefing between the Captain and the In-charge cabin crew member at the beginning of flight. Although guidelines for such minimum briefings are in place and are supported by regulatory requirements in some countries, the pressures of the operation coupled with frequent crew changes on short haul routes, severely affect the quality of exchange at these times.

Cooperation between these two groups is frequently based on the cabin crew members' perceived or real degree of willingness to provide basic service to the flight crew during flight. Research confirms (Chute, 1995) barriers to cockpit/cabin communication include flight crews treating cabin crews disrespectful. Over 36 percent of the respondents complained about flight crews rude, aloof, demanding behaviour and boarding the aircraft late. A total of 12 percent objected to being ignored, briefing the In-charge cabin crew member only, or not receiving a briefing at all.

Flight crews on the other hand complained about being ignored as 'defined by cabin crews hiding in the back of the airplane, snubbing them, not checking on them, not visiting the cockpit, and not bringing them food or beverages'.

The range of expectations from the flight crew can vary substantially. The author personally witnessed cases where demands reached unreasonable levels. Although these cases are not typical, they do deter from the respect that needs to be nurtured for overall positive work relationships. There is the example of the commander who insists on being served first class food prior to the service having commenced to the paying customer. Some extreme cases such as bringing one's own teapot and insisting on being served hot beverages in first class china cups despite company policy, are another. I have been asked on one such an occasion to intervene with a flight crew who requested that beverages be provided for their layover, only to face an incredulous response when reminding them that this would be considered theft. Attitudes of entitlement are always counterproductive; they have no place on or off the aircraft. Respect has to be earned in each situation, no matter how insignificant it appears to be. This goes equally for the cabin crew and the In-charge. Some cabin crew members have erroneously interpreted the 'power to act' syndrome. In-charge cabin crew members, I interviewed, reported difficulties in curtailing co-workers who believe they can leave their workstations at any time and enter the cockpit without requesting permission via the In-charge crew member.

Crew coordination and maintaining formal communication processes are more difficult to sustain on large aircraft with multiple cabins and decks and with cabin crew teams composed of more than five or six members.

Cultural Differences in Communication

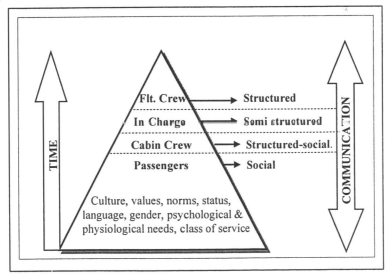

Figure 4.3   **Communication Style Requirements**

Figure 4.3 illustrates the communication style requirements between the flight crew, the In-charge, and the cabin crew members, including the predominant communication style used by passengers.

The top of the pyramid represents the strict communication norms pilots must follow to ensure the safe operation of the aircraft. These formalized communication processes are evidenced in the standards set for pilot to pilot and pilot to air traffic controllers communication. Messages cannot be ambiguous, and when in doubt, messages will be verified. The intensive training of focusing communication on technical problem solving, reinforced on each flight operated, affects flight crew expectations about their communication style with the cabin crew.

Cabin crew members on the other hand are not selected or specifically trained to communicate with the flight crew in a similar manner. Although In-charge cabin crew members do receive special training to varying degree depending on the airline, these skills are not reinforced with the rest of the cabin crew (Chute & Wiener, 1995).

As time pressures increase (left arrow pointing up), so does the need for precise and structured communication (right up-down arrow). Unless these skills are developed and honed in the working environment, effective and timely problem solving is jeopardized. With the growing trend of multi-cultural and multi-lingual crew members, the risk of ineffectual crew interaction increases. Cultural difference training is one strategy to address this issue.

In the work environment 'cabin crew terminology' and sign language facilitate their task interaction, especially when service delivery competes with flying time. The air traveler can observe this when cabin crew members working from their trolleys signal their co-workers for additional supplies.

This practice is not applied in other situations. The majority of cabin crew members practice a communication style rooted in social dialogue prevalent to their own cultural background and norms established in each team. Although this may suffice for dealing with relaxed social intercourse with passengers, it is not conducive to effective problem solving. This is supported by our research identifying the high percentage of incidents related to compliance issues. As previously discussed in Chapter 2, cabin crew members feel generally uncomfortable dealing with a passengers reluctant to comply with safety regulations or policies established by an airline that are contrary to their expectations. Incidents involving status conscious individuals, who are used to enjoying special privileges, receiving special treatment and making their own rules, are more likely than others to object to an authoritarian approach by the cabin crew. This reality should be addressed in the training provided to cabin crews at large to provide them with skills related to social etiquette in an effort to protect them from unnecessary conflict and verbal abuse.

Following are some examples where inadequate crew communication and intervention has exaggerated a situation or delayed its timely resolution.

Case A - This incident occurred on a light transport aircraft:

Had an ex-crew scheduler for our regional airline (now employed in a similar capacity by the code-sharing partner) on board. Not a certified airman! He asked Flight Attendant to visit cockpit en-route and without checking with me, the dumb s___ Flight Attendant opened the cockpit door and let him into my cockpit!! I asked non-revenue passenger if he was a licensed airman and he said no. I politely told him that the cockpit was off limits to non-certificated personnel. He left. Flight attendant manual has listing of who is allowed into the cockpit and procedures to be used to allow people into cockpit. The flight attendant simply didn't think, and as a result, put me and my First Officer into a violation situation. (ASRS report #236191)

Case B - The following passenger complaint illustrates ground staff and cabin crews' inappropriate handling of intoxicated passengers on a charter flight:

...My wife and I were looking forward to a wonderful holiday. This was my wife's first time to Honolulu and she was very excited. As we were settling into our seats, we noticed at least four to five people who were drunk and let on to the aircraft. Before take off the head flight attendant notified them that they would not be served alcohol during the flight

Seated in the aisle directly in front of us was a gentleman who was not intoxicated at the time of boarding and seated behind us in the window seat was a gentleman who was intoxicated before we left Vancouver. They started to order drinks. As time went on more drinks were ordered and these men were served every time. They became loud and started to yell directly above our heads. Another man seated two rows in front of us and directly in front of the man in the aisle seat stood up and started to yell at these two men to sit down and shut up. Of course a verbal fight started. The flight attendants came and told all involved to sit down and be quiet. Drinks were still being served to these men. Things settled down for a few minutes and then they started to yell again. I decided not to say anything as things were way out of hand, and the flight attendants were trying to stop the arguments and calm everyone down. My wife counted at least ten beers being served to the man directly in front of us who became intoxicated. The man in the seat beside him tried to calm him down with no luck, and was told by the head flight attendant to stay out of it.

In the last half hour of the flight, the head flight attendant had the man in front of us in the aisle seat move to a seat at the front of the airplane, and things got quiet. This whole incident was handled in a very poor manner by the flight attendant, and alcohol should have been stopped before this situation arose.

Case C - The following incident occurred on a B747 transatlantic flight and involved a group of twenty family members ranging in ages from three to sixty five. The events are summarized as follows:

This group boarded just minutes prior to departure. Although they showed signs of disorientation, they did not appear intoxicated. They were seated in the rear economy cabin and started to be unruly and disruptive right during taxi by standing up and roaming up and down the aisle. The cabin crew serving beverages noticed that miniatures were missing and alerted the coordinator who had attempted to obtain compliance from the group members all along. One of the male members of the group was particularly confrontational. Approximately 1½ hours into the flight, the In-charge cabin crew was notified of the persistent difficulties the cabin crew had with these passengers. After the In-charge went to intervene and advised these passengers that no alcohol beverages would be served to them and informed the other cabin crew members to ensure that this would be the case. The In-charge then proceeded to inform the captain of the situation.

As the flight progressed, disturbances continued, escalating to smoking in the washroom, spitting on other passengers, shouting verbal abuse and throwing food. The captain was called to intervene on several occasions, including to break up a physical fight between two members of the group. In consultation with the In-charge, it was decided to move the instigator to an unoccupied area in business class where three other male passengers kept him subdued and monitored him for the rest of the flight.

At one point the In-charge had to ask other cabin crew members who had been drawn to the disturbance area to return to their assigned work areas.

Several cabin crew members objected to the decision taken by the captain and the In-charge, causing major dissention in the team with the In-charges attention now being split between upset co-workers, the out of control individual and upset passengers being witness to the upheavals on several fronts. One cabin crew member went as far as leaving her station without permission of the In-charge to voice her opposition to the captain for some of decisions taken.

Although the captain had considered diverting the flight, and having the enraged passenger taken off, time remaining was such that he decided to proceed to their final destination in order to avoid further inconvenience to the remaining passengers.

In summary, this incident demonstrates the need for improvements in a number of areas: screening of customers prior to boarding, clarification of roles and responsibilities regarding denied boarding of disruptive passengers, cabin crew communication with the In-charge and the management of unruly groups, standardizing terminology to identify level of severity when confronted with a threatening passenger situation, communication with the flight crew, and flight crew training concerning their responses to escalating passenger conflict situations.

The airline on whose flight this incident occurred initiated a number of important changes in their policies and procedures as well as in their training. The active participation of concerned management in lobbying IATA was part of a change in airline management approaches to such incidents as reflected in the IATA guidelines to its members. The last case served as a 'Tipping Point', a term coined by author Malcolm Gladwell. He proposes that social phenomena, in our case the fascination with 'Air Rage' are a sort of virus that flourishes under certain conditions and becomes epidemic.

Safety and Service – the Central Paradox

Old images and terms have a strong hold in the public at large. Even today in the year 2000, both members of the public, the media and the aviation community itself, still refer occasionally to female cabin crew members as stewardesses. This is not surprising, given that passengers experience cabin crews predominantly in their service role.

In performing safety duties routinely, complacency sets in. This goes for cabin crews and passengers alike. Safety videos instead of live safety demonstrations are commonplace on major airlines. The quality varies, holding limited interest for the travel and media savvy passenger.

Although safety videos improve the consistency of the message and free the cabin crew members from repetitive duties, this method may occasionally reduce respect for the cabin crew. This happens when professional appearance and demeanour is out of sync with that portrayed in the video. The video negates an important opportunity for establishing an unambiguous rapport with the passengers on the basis of safety related duties. Primary communication on an important aspect of cabin crew members' role is taken out of the personal realm and delegated to technology.

Marketing departments are constantly searching for new ways to keep customer loyalty and attract new ones. The competition is fierce and resulting in decisions that are often difficult to reconcile with the operational environment, the time pressures and the safety responsibilities of cabin crew members. More is not necessarily better when the cabin crew is unable to deliver consistently what has been promised to the customer.

The cabin crew, like no other employee group in this system, is caught between these opposing forces. On short-haul to medium flights, service and safety tasks require military precision for getting the job done on time. With high load factors and varying cabin crew staffing for service purposes, the risk increases dramatically for passenger malcontent.

A passenger identifying a problem to the cabin crew under these circumstances is not likely to get any satisfaction because of time constraints and a stressed crew. Some cabin crews explain that reverting to an autocratic communication style is the only feasible way to maintain control: 'Operational pressures don't allow for lengthy discussion; we have a job to do.'

Shifting Control

Flight crews are primarily concerned with operational safety. That is the historical foundation of their role. Based on military and marine traditions, the captain in command determines what risks to take and assumes full responsibility for the welfare of all persons on board. Additional crew members and cabin crews are bound in this classic pyramid structured organization from take-off to landing. The captain's decisions are final.

Despite significant technological changes in the cockpit affecting the traditional interface between man and machine, and despite the concept of crew resource management or any derivative thereof, the hierarchy of command is still in place and for good reasons.

The cabin crew organization is based on the same hierarchy structure with the 'number one' (or lead flight attendant, In-charge or Customer Service Director or any other name designating this rank), given the highest authority as the leader of the team.

Notwithstanding what for years was an accepted reality of a rigid hierarchy structure in the airline industry, authority and interpersonal issues have changed as much within the cabin crew workforce they have in society at large. The history of cabin crew labour associations and their role in the labour movement provides an excellent insight into the some of the most dramatic developments affecting relations with the customer, their members, the airlines, regulatory authorities and politicians.

With Human Rights legislation in place, leadership positions in the passenger cabin also had to open up to women who previously were prevented from assuming these positions. Like any major change, it did not come without lasting bitter feelings from their male colleagues that affected interpersonal relations for years to come.

In today's work environment competent cabin crew members will resist the command and control of any insecure leader because they believe they can function well without them.

Those of us in management who experienced the desperate search for solutions in the turbulent 1980s and continuing into the 1990s, are very familiar with the flavour of the month, how-to recipes our companies embarked on to cope with the changed business environment during the recession. Top management approved of campaigns such as organizational excellence, total quality management, customer care, 'Moments of Truth', employee participation and empowerment, which were destined to be expensive cosmetic exercises. At the heart of these failures were rigid hierarchical management structures and managers who spouted the latest management jargon but whose actions demoralized the workforces with failed promises. At the height of organizational restructuring at one Canadian carrier between 1989 and 1993, implementation of new management theories were as predictable as quarterly statements.

Frank Lorenzo, former head of Texas Air embodies a management style that Charles Garfield defines as the lone pioneer style. After purchasing Eastern Airlines, Lorenzo, convinced that dramatically lowering labour costs were the only salvation, refused to engage in discussion with Charles Bryan, head of the Machinists union and opted to dismember the airline instead. At Continental Airlines, he used bankruptcy to break the unions.

His leadership was admired and had significant influence on other airline leaders at a time when the recession and deregulation in North America added to enormous financial losses. Behind all the new management theories and related jargon was a blunt reality. It was an era of bean counters. Massive layoffs made headlines almost weekly. Management by intimidation reigned supreme. With no protection from an association, middle and line management was systematically dehumanized. The predominant concern at all levels of the organization except of those of the 'inner circle' was survival. The organizational stress ravaged the workplace like the black plague. Panic was in the air. People had difficulties concentrating on the job at hand. Service cuts prompted greater numbers of customer complaints. Cabin snags reported in the aircraft logbooks were repeatedly ignored, causing cabin crews to have to deal with increased numbers of passenger complaints. Absenteeism increased together with employee injuries. There was a measurable correlation between the failing financial health of the corporations and these performance areas.

Ultimately this 'lean and mean' approach did achieve some measurable results as evidenced by the corporate balance sheets but has left employees who remember these years, weary and cautious.

Organizational Issues

Flight crews and cabin crews are usually organized under different departments, with flight crews reporting to Flight Operations while cabin crews report to In-flight Services as one of the marketing divisions. This division is a clear indication of the primary value of these groups to the organization. Flight Operations is predominantly concerned with safety while cabin crews are viewed as a predominant marketing tool. A study at two American carriers indicates that approximately 63 percent of flight crews and 68 percent of cabin crew members surveyed confirmed their preference to report to the same department (Chute & Wiener, 1995). Clearly this would harmonize safety related objectives, procedures, training and communication between these two groups.

Recommendations relative to an organizational realignment of cabin crews to Flight Operations do not meet the airlines' internal interests. In-flight Services departments are often considered training grounds for fast track airline executives or convenient retirement positions. In addition balance of powers and influence are also considerations to keeping the two groups separate. Given the fact that flight operations executives most often are qualified pilots with their seniority intact throughout their tenure, a case could be made for conflict of interest during negotiations. Already there is sufficient evidence in labour negotiations that cabin crew labour associations follow the principles of a 'me too' bargaining position that would further challenge management negotiators should both groups be united under one department head.

In essence, management's concerns appear to be self-serving. Public pronouncements of 'safety first' must be understood as good public relations, however a closer look at organizational structure reveals internal sensitivities in assigning different value to the contribution of flight and cabin crews.

Terminology, Communication, Role Conflict

The aviation system uses its own terminology to communicate internally, thus creating a special language alien to both insiders and outsiders. This has been well documented in research on the effects of a deficient common language for cabin crews to identify safety-critical issues (Chute & Wiener, 1994; 1995). When this carries over to communications with the passenger, the passenger can feel excluded when dialogue and understanding between the public and the crew is important.

Based on independent research conducted in the U.S.A, Canada and Germany between 1992 and 1997, passengers do not feel airlines effectively communicate with them. Typical complaints are that airlines are ignoring increased awareness and concern about passenger safety and quality service. The respondents shared the view that airlines and their employees fail to provide sufficient safety-related information in order to achieve a more cooperative relationship with their customers.

Cabin crews are expected to enforce safety regulations, to recognize individual customer needs and take responsibility to ensure that they are satisfied. This role of both enforcer and service provider causes at times conflict and confusion. Some flight attendants express their opinion, backed by regulatory regulations that 'they are here for safety reasons.' This implies, 'I am not here for service.'

The relationship established between flight attendants and their customers can make such a difference. Passengers love naturalness and covet the original performer who also knows when to stop. Knowing this, some airlines search out these characteristics, and build their reputation on their friendly service. Cabin crew members tell jokes, and deliver a lighthearted version of the mandatory safety demonstrations.

The introduction of high tech service products in the passenger cabin reflects the products available to the consumer at large. They are designed to entertain and increase passenger control over their immediate environment. Music, movies, video games, on-board telephones, ports for laptop computers and sophisticated electronic manipulation of seat in premium class change the role of cabin crews from one where personal interaction with the customer during service down-times was commonplace to one of greater remoteness.

This development also reduced the opportunity for cabin crews to interact on a more personal and relaxed level with the customer, especially on long-haul flights. Another consequence is the reduced ability to spot potential problems and connect with passengers, and getting to know them a little. The pay-off for cabin crews in solving small problems, in giving a sincere response to a question, or simply to create better understanding, was an opportunity to establish trust and gain respect for their role.

Similar to the developments in the service industry, the beginning of civil aviation passenger service was 'high touch'. Currently, customers experience increasingly 'high tech' service as we discussed in Chapter 3. Quality service experts believe that the third phase of service will be 'high touch through high tech' (Berry, 1995).[7]

If this is so, cabin crews need better access to airlines' automated information systems in order to fulfill this role. The technology is there, however cabin crews have generally less direct access to pertinent information compared to flight crews and their colleagues in reservation, sales or marketing. Despite being with the customer the longest, they are generally dependent on other colleagues to provide them with pertinent passenger information. Consequently, they are left at a significant disadvantage when interacting with the passenger. This illustrates a critical weakness in the thinking of airline management. Instead of affording effective use of technology to cabin crews, they continue to be last in line when corporate decisions are made regarding who in the organization is best served by this technology.

A New Orientation

Few unionized workers seriously consider promotion into the airline management ranks as desirable. The reasons include uncontrolled working conditions, loss of pay to hours worked, loss of seniority, (a factor that drives working conditions on the line), insufficient compensation or no compensation for overtime, no guaranteed days off, unstable management organization, lack of upper management support, uncertainty, and a lack of job security.

However this situation does not apply to organizations where a new trend in orientation is evidenced by greater emphasis on employees' social needs with features such as flexible benefit programs and policies, suited to the individual life style, incentive bonuses, job sharing, cross-functional teams in product development, and access to information, previously reserved for top management.

This trend has transformed many companies, however the creative talents of the unionized airline workers have not been as successfully utilized as those of some of their counterparts in other industries. As for management, the workaholic environment and its pace continues to this day. Quantitative objectives and measurements permeate aviation.

One such example is America West Airlines, now a major carrier based in Phoenix, Arizona. Without the burdens associated with mature organizations, its founder, chairman and CEO Ed Beauvais launched America West in 1983. The company became a success story unparalleled in aviation and in the deregulated US airline industry. In 1989, only six years after its formation, the U.S. Department of Transportation declared the carrier a major airline with annual revenues of over US$1 billion. America West is a low- fare full service airline, and has expanded their destinations to Canada and Mexico.

Recognizing the need for strong relationships in order to succeed in the profit driven aviation industry, America West was also first in entering a code sharing agreement with another American carrier, Continental Airlines. This was a significant move and set a new trend for other airlines to follow. Apart from a keen pursuit of bottom line economics, relationship management with other corporations extended to its employees who are treated as full partners in the enterprise.

Commitment to employee partnership is reflected in the employee stockholder plan that is enhanced by a profit sharing and incentive stock options program. An open door policy coupled with an 800 number hotline gave employees full access to communication with the executives.

Still a fledgling airline, America West instituted a progressive childcare program that received national recognition. With changes in society, the increase in single parent families places significant stresses on workers that have concerns regarding access to safe childcare. America West recognized this issue and responded in a uniquely effective way. This and other programs aimed at helping employees in difficulties and rewarding them for exceptional performance shows a broad based commitment towards treating employees as full partners.

Few airlines can show similar success in their relationship management with their employees. The long-term success of these programs in a mature organization however is difficult to maintain. Changes in executive management and leadership styles have a major impact on organizational culture.

WestJet, an upstart in the Canadian aviation industry has modeled itself after America West. Now in its sixth year of operation, it is expanding its route network into Eastern Canada, adding to its fleet and enjoying growing success.

In July 1999 it completed its Initial Public Offering of 2.5 million common shares at C$10.00. Nine months later it announced a stock split based on investor confidence with a share value that had more than doubled.

The restructuring of the Canadian airline industry with the merger of Air

Canada and ailing rival Canadian Airlines International in 1999 provided an opportune environment for expansion plans. This marks a fresh cornerstone in the airline marketplace with Eastern destinations providing Canadians with a new choice for low-fare, short haul travel.

Combating Disruptive Passenger Incidents

Airlines have a legal duty to ensure a cabin environment free of unlawful interference with crewmembers, violence and abuse based on general principles of laws, including those governing the protection and compensation of employees. In North America, much emphasis has been placed on legislative issues over the last four years, especially by labor. Far less emphasis has been placed on organizational policies and practices and the importance on gathering relevant data.

In-house experiences with disruptive passenger responses are invaluable in building into training programs. On-board defense strategies proven to have been successful help develop crew confidence and competence.

Major airlines have adopted a no-tolerance policy towards abusive and violent passengers. Many others have not. The competitive nature of the business combined with weak legislation is favouring such neglect to the detriment of social peace and employee safety in the passenger cabin.

Training programs adopted by some airlines to address passenger conflict and violence vary. Delta Airlines for example, uses a program offered by the Verbal Judo Institute teaching varies communication techniques designed to prevent situations from escalating.[8] Airlines already carrying restraining devices, are required by law to train crews in the proper use of 'restraining' ties, including legal issues. The emphasis was in essence reactionary after an incident had escalated into physical assault.

UK-based Securicare offers 'Conflict Management Training' to airlines. The two-day course offers theoretical and practical training with a focus on the environmental constraints of the aircraft cabin.[9] Of course many airlines develop their own in-house training syllabi following the renewed interest in these topics. I recall an extensive training program with one major Canadian air carrier in the early 1970s when conflict resolution and problem solving skills were taught during induction training, followed by further specialized training for In-charge cabin crew members.

The effectiveness of such training however has to be measured in the operational work environment and reinforced with coaching and in recurrent training to take effect in the long-term. This is where regulatory requirements as an intervention can play an important part to ensure these training programs are more than a one-shot deal. The many cabin crew members I interviewed, report different reactions to the various training programs provided by their airlines, and support this observation. Without regular refresher training, complacency sets in, since the change in behaviour cannot be expected to take place without such support.

Celebrating Success

Airline workers who bring difficult passenger incidents to successful conclusion deserve to be celebrated. They need to be recognized by their airline and thanked for their accomplishment. Crews faced with a situation such as the incident on Alaska Airlines on March 16, 2000, involving a male passenger storming through the locked cockpit door and attacking the First Officer in an attempt to reach for the controls, are one such example. A dinner was given in their honour, and together with the passengers who assisted the crew in subduing the attacker, they were presented with the 'Above and Beyond Award.[10] Recognition, even for minor incidents, goes a long way. It is incumbent on those who witness these situations, a colleague, the In-charge, the flight crew, to immediately thank and praise the ones involved.

Many airlines have already systems in place to collect such information; they have to be activated to ensure management is aware and do their part in reinforcing these positive efforts. Publicizing the events serves both as a learning opportunity for others to see how different conflict situations have been resolved, and as an affirmation of the organization taking pride in their workers' accomplishments.

Conclusions

I have attempted to reveal the connection between the historical evolution of the role of cabin crew, organizational factors, including those on-board, and cultural aspects playing an important part in managing social peace in the passenger cabin.

The current barriers do not have to persist. It too requires teamwork. Although there is no quick fix to improving this situation, regulatory authorities as well as airline and labour management have a primary responsibility to act on these issues.

Regulatory Authorities have the responsibility to ensure that clear training standards are embedded in their regulations in support of fostering on-board teamwork, communication and coordination. Unfortunately, this area is weak in most cases, except as evidenced by the Transport Canada Training Standards that clearly spell out what must be covered in all phases of training.

Airline organizations have control over a broad range of strategies to best support the effective achievement of their safety and service roles. These range from defining hiring criteria to training to aligning organizational structures. The following questions should be addressed in optimizing their efforts:

1. Is our organizational structure optimized to fully integrate safety and service goals?

2. Do we actively pursue teamwork in this company? Do we have clear safety guidelines for marketing to use in the design of products and services? Is marketing validating its product design ideas with input from Flight Operations and In-flight Services? Is teamwork viewed as a cultural imperative?

3. Do we hire cabin crew members whose skills match current and future requirements for on-board problem solving and conflict resolution?

4. Do we have an integrated conflict management program?

5. Do we make the needed efforts to break down organizational barriers to teamwork and communication? Do we recognize the conflicting messages we give cabin crew members in respect to their roles and responsibilities?

6. Does our internal communication play an active part in narrowing divisions between Marketing, Flight Operations and In-flight Service? Are we listening and mediating their concerns?

7. Do we have a process in place to discuss mutual expectations and perceptions on a regular basis?

8. Do we reward and celebrate teamwork, positive outcomes of passenger incidents? Do our people believe collaboration leads to better results?

9. Do we provide adequate briefing time for crews?

10. Do we teach CRM in joint training sessions? What other methods can we use to create better understanding between flight and cabin crews?

11. Do we teach teamwork throughout the organization?

Notes

[1] Chute (1995) and Wiener (1988) demonstrate that the change from a three flight crew cockpit to the two-flight crew cockpit has affected the communication and coordination between the flight and cabin crews.

[2] Dunbar, Chute and Jordan (1997) demonstrate cabin crews are not 'technically aware nor articulate in order to facilitate effective information transfer'. Although the focus of this research was aimed at cabin crews' ability for detecting anomalies relative to operational safety, the general lack of training in this area also impacts on effective communication with passengers expressing concerns or fear of flying. The lack of technical training for cabin crews has already been identified as an issue in a number of accident reports (National Transportation Safety Board, 1992; Transportation Safety Board of Canada, 1995; Chute and Wiener, 1996).

[3] *Reuters*, 23 June 2000.

[4] *United Airlines Press Release*, 23 June 2000.

[5] *Las Vegas Review Journal*, 29 September 2000.

[6] *The Edmonton Journal*, 20 September 2000.

[7] Berry, L.L. (1995) 'On Great Service – A Framework for Action', The Free Press, New York., pp. 158-159.

[8] *Star-Telegram*, 6 July 2000.

[9] Prew, S. (1999), 'Training to Combat Air Rage', in CAT Magazine, The Journal for Civil Aviaition Training, June, pp. 34-39.

[10] *Airlinebiz*, 8 June 2000.

5 The Aviation System

Introduction

This chapter examines organizational and cultural factors involved in the phenomenon of passenger misconduct. The aviation industry has only recently recognized passenger misconduct as a threat to aviation safety. Although a significant body of research into the impact of organizational cultures on safety[1] has been produced over the last couple of decades, little has been done to examine the link between organizational factors and passenger misconduct.

This chapter looks at passenger misconduct as a type of 'relationship accident' that in the majority of cases could have been mitigated by the aviation system and its operators. The goal is not a full-scale analysis of all organizations directly or indirectly involved with the prevention and management of passenger misconduct.

The Airlines

The essence of the airline business is to provide safe and cost effective transportation for people and goods. The airlines' relations with their customers are based on service principles while operations including Maintenance functions follow in essence production principles. These principles are not easily reconciled. The organizational structures reflect the increasing complexity of business, however the core functions of Operations and Marketing continue to be critical for the airlines' success. Customers' primary concern with schedules, routes, connections and fares has evolved to now strongly focus on service issues. A new relationship with the airlines is the goal of the passenger rights movement. Safety and maintenance records, although rated as very important by the public affect purchasing patterns less than service, and looming labour unrest.

Historic events transformed the early aviation culture from serving the militarization of World War I to civilian and mail transport in the 1920's and 1930's to transport of troops and medical supplies in World War II. Much experience was gained and fuelled the need for better airplanes to fulfill their diverse missions according to Industry demands. Engineering milestones in the early 1950's came with the introduction of the turbo-compound piston, followed by turbo-props in the late 1950's. The 1960's brought on the 'true jet' age. Air travel became faster and more affordable for a greater sector of the population.

With the Boeing 747 Jumbo jet released into service 30 years ago, the new era of mass air travel took flight. Since then, 1,238 B747's have been delivered. Its competitor Airbus Industries now threatens Boeing's profitable monopoly position. Airbus Integrated Co., as it named coinciding with the announcement of a leaner corporate structure in June 2000, also gave the green light to sell its A3XX, a new generation large aircraft.[2] Although Airbus claims that the A3XX would be 17 percent more cost efficient than the B747, and some airlines have signed letters of intent, Airbus needs an undisclosed minimum of orders to commence production.[3]

What once was the exclusive means of transportation for the financially privileged had become 'an equal opportunity' means of transportation for all sectors of society. Clearly, the democratization of air travel has generated a complex system to the detriment of the more conventional means of transportation such as trains. The impact of the expanding aviation system is far reaching; national economic prosperity is linked to what occurs in the industry. The consumer's experience is one aspect, however the need for support functions and the resulting job creation are major. In the example of the O'Hare Midway airport system, a study commissioned by the Midwest Aviation Coalition concluded that more than U.S.$35 billion annually were injected in the region's economy while supporting more than 500,000 jobs.[4]

Serving the Public Interest

Airline business needs are by necessity linked to government relations. Evolving from its primary mission to serve national strategic interests in WWI and WWII, (Williams, 1993) the debate surrounding deregulation in North America and subsequently in Europe became intense. Political considerations include a benign regulatory environment, adequate funding for the maintenance and expansion of a viable infra structure and support for beneficial bilateral agreements. The needs apply to all airlines, regardless of their ownership status or type of operation.

In recognition of mounting criticism regarding the current difficulties coping with the existing volume of air traffic, the U.S. in its budget proposal released for the fiscal year 2001, announced an increase of 6 percent for the FAA, bringing the total budget for aviation safety to U.S.$ 11.3 billion.[5] FAA Administrator Jane Garvey announced aggressive modernization of the airspace system infra-structure needed to support the strongest growth sector of the industry predicted in the regional jet fleet. The FAA forecast, released early in 2000, predicts an annual growth rate of 13.4 percent through 2011. The FAA already opened a new air traffic control facility in the Baltimore-Washington region to alleviate the growing problems that come from growth.[6]

State owned airlines focus on protecting the status quo. Performance pressures are absent when subsidies are readily available. Privately and publicly owned airlines focus on competition, shareholder value and financial performance with varying degrees of a customer-focused approach to business. Because of the Marketing departments playing a key role in developing strategies to ensure the profitability of their companies, the values of safety versus service are not fully aligned. Public interest is not necessarily researched by Marketing with the view of the impact on safety or social consequences in the passenger cabin.

State ownership versus private ownership raises questions as to how one serves the public interest better than the other. Each time an airline moved from being State owned to privatization, one of the central issues focused on public interest. Airline mergers equally inspire heated debate. The example of Air Canada demonstrates that these concerns are valid. In a comparison of airfares between ten major Canadian cities supports consumer complaints that fares are lower in those areas where competition exists, and has resulted in higher fares where there is none.[7] Monopolies of any kind will unleash criticism that competition is at risk and the public's interest jeopardized.

During the transition of integrating Canadian Airlines, complaints have risen sharply. Public pressure resulted not only in the federal government appointment of Air Travel Complaints Commissioner, but Air Canada followed suit with appointing its own ombudsman, believed to be the first of its kind in the industry. The position reports directly to the president and is responsible to liaise with the Canadian Transportation Agency, the government agency charged with administering transportation legislation and policies.[8]

The Pioneering Spirit

Aviation history is built on pioneering into the unknown, technological revolution, linking diverse civilizations through a transportation net equal to none. Bush pilots, fighter pilots and engineers were the early captains of this new industry. These leaders set the trend for an organizational culture that has rewarded autocratic decision-making in a firmly established hierarchical structure until the 1980s. Organizational charts of that time reflect this reality. Flight Operations reined supreme, being at the top of the organization with a span of control across all others. In line with societal values and norms, authority figures commanded respect, and the consequences for challenging authority in any form, were severe. Aviation still considers itself to be the leader in transportation despite evidence to the contrary.

A Look at History

Governments controlled and regulated the airline industry since its beginnings in the 1920s, deciding on a system of contracts for mail carriage thereby guaranteeing the development of such services in a stable financial environment. In the United States, it was the Contract Mail Act of 1925, with amendments to the Act in 1930 with further changes introduced by the Air Mail Act of 1934. The result of this Act was the creation of three separate regulatory bodies as part of a highly bureaucratic system (Williams, 1993).

The Civil Aeronautics Authority (CAA) was created in 1938, following a series of fatal aircraft accidents that highlighted the need for an organization solely dedicated to air transport. Its powers were sweeping, ranging from price regulation to mail rates to entry of airlines on interstate routes, and control of all aspects of safety.

Similar developments occurred in Canada. In 1927, the Post Office started to award official contracts for carrying mail. In 1936, the Department of Transport (DOT) was established to develop a unified transportation system that was capable to support a national mail service, and the war effort. The DOT now combined departments of Railways and Canals, Marine, and the civil aviation branch of the Department of National Defence. Its immediate challenges were the building of airports, harbors for the naval fleet and a highway system to open up the country in new ways.

Airmail had already been carried between London and Paris since 1919, while transcontinental mail service to the United States was fully established by 1925. A basic infra structure to safeguard these services was already in place. Radio beacons assisted flight crews to keep on-course; national weather offices provided information to plan their flights safely, and airport lights permitted night flying. Canada had none of these services in place at the time it awarded its first mail contracts in 1927.

The creation of a government owned airline in 1937, Trans Canada Airlines had the mandate to be the official carrier of mail and passengers. Canada's prosperity was clearly dependent on an efficient transportation system. Trains and independent small operators were the main providers of mail and passenger transport, no longer sufficient to compete with the more established air industries in the United States, Britain and Europe.

Because of the occurrence of a number of fatal aircraft crashes, some linked to competing operators overloading planes, the need for the safe operation of aircraft became a driving force in establishing air regulations. This also extended to passenger safety regulations and the emerging safety role of cabin crews.

Initially, Canadian aviators viewed the carriage of passengers as a sort of nuisance while they saw their primary duty in delivering the mail. Don MacLaren, a war ace and one of the pioneering pilots on these Canadian mail runs, recalls:

> Passengers sometimes wanted to fly with us – I could never see why they did – and why they wanted to pay for the ride. Some fortified themselves with the stuff people use to fortify themselves with and you wondered whether they flew because they had been drinking or drank

because they wanted to fly. After watching one passenger trying to open the cabin door to climb out on to the wing at 5,000 feet one night, we decided to look the passengers over more carefully before boarding.

Even the Post Office attempted to bar passengers at one point, complaining that passenger carriage slowed down the mail delivery schedule. Previously, flight crews had been taken greater risks in flying under more hazardous weather conditions. The added responsibility for human carriage changed their attitudes towards continuing to take such risks.

In the example of Canadian mail service, the first airplanes used were the Fokker F-14s and the Boeing 40B. The flight crew sat in a raised open cockpit behind the passenger cabins. The purpose of this was to provide the flight crew with an increased field of vision, aimed at improving their chances for survival in case of an accident. The Post office was concerned with passengers tampering with the mail that was stowed in the washroom, and eventually mandated the aircrafts to be modified with a locked compartment beneath the cockpit.

These developments point to an initial culture within the budding air industry that favoured cargo over the carriage of passengers. Captain MacLaren's assessment of passengers, humorous as it may seem, has a new relevance in today's environment where alcohol abuse is still a major factor in on-board incidents.

Deregulation

The Civil Aeronautics Authority in the United States was faced with a changed situation after the Second World War, when new charter operators entered the market, competing with the established air carriers. They offered substantially lower fares, prompting the Civil Aeronautics Board as the CAA was now called, to protect the scheduled carriers by introducing limits on the number of flights the charter operators could undertake and grant fare increases when industry profits were low. Market conditions were not taken into account while decreased load factors were blamed as a result of scheduling rivalry (Williams; 1993).

The span of regulatory control did not extend to the California interstate operations. This allowed airlines serving this market to engage in price competition based on a keen focus on operating efficiency. Economist of the early 1970s generally agreed that the profitability of Pacific Southwest Airlines (PSA) was in large part due to the lack of regulatory interference by the CAB.

Researchers continued to compile evidence that the CAB's continued protectionism of airlines was not in the public interest. Forces for change were strong and resulted in the Airline Deregulation Act in 1978. The impact of this Act was dramatic and sweeping. Airline executives at that time had no experience with survival strategies now necessary to compete in the deregulated environment (Williams; 1993).

Airline strategists had to abandon early route expansion strategies in favour of developing a new set to respond to increased competition. Williams (1993) summarizes these under two major options: Defensive tactics employing cost reducing activities, and revenue generating activities to operate within the new commercial environment, and offensive tactics by building barriers with the aim to transforming the new competitive environment.

This new reality provided new opportunities for marketing professionals in a changed airline industry. They were the people with the expertise to ensure survival and profitability. Powers shifted as a result from previously predominantly operational driven organizations to marketing driven airline organizations.

In Canada, the situation was similar with the Canadian Federal Government transforming the existing duopoly between the Crown Corporation Trans Canada Airlines, by then Air Canada, and the privately owned Canadian Pacific Airlines in the late 1970s. Deregulation in the United States also affected the Canadian electorate to pressure the Government for changes to its regulatory policies to encourage competition. During the period of 1979 to 1988 a number of major Government initiatives culminated in The National Transportation Act of 1987 to become law in 1988. Canada was divided into North and South regions, with the latter enjoying few restrictions on pricing and market exit.

Neck-to-neck competition between both airlines ensued with the addition of Wardair allowed to operate domestic advanced booking charters. One of the many marketing decisions aimed at enticing customer loyalty resulted in the offering of free alcohol to all classes of service on designated domestic flights. The author recalls the results of the first B747 flight on the Vancouver-Toronto route introducing this service with the average consumption of 2.8 drinks per passenger, and the bar practically depleted upon arrival.

Cabin crews were distressed. Passengers, enticed by advertising, felt entitled to consume or pocket as many miniatures as they could, with the cabin crew being quite unprepared to deal with this situation. Clearly, the impact on the social atmosphere on-board had not been a serious consideration for Marketing.

Aviation System

Each country has its own organizations which although similar are uniquely matched to their own experiences in aviation development. Relationship styles between these organizations differ from country to country.

The organizations are charged with different levels of authority and and ultimately impact on how the traveling public receives its service.

Investigation boards play an important part in the overall aviation system although they are not empowered to see these recommendations implemented. It is up to the regulatory authorities to establish the need for new regulations that would legally oblige airlines to implement these regulations in all aspects of their operation. One such example is the recommendation by the National Transportation Safety Board (NTSB) for a ground movement safety system that will prevent runway incursions.[9] The implementation of such safety recommendations is not possible without major financial commitment from governments. With growing consumer advocacy, some of the issues will be receiving greater urgency than others for legislation to be passed. Of course there are cases where an airline involved in a fatal aircraft accidents has taken immediate remedial steps to prevent a similar occurrence by implementing a number of recommendations in advance of changed regulations. Such was the case with Air Canada following the investigation and recommendations after the Cincinnati incident.

The Canadian Transportation Safety Board (TSB) highlighted risk of mid-air collisions and runway incursions in its annual report released in July 2000.[10]

International Civil Aviation Organization (ICAO)

The International Civil Aviation Organization (ICAO) is a specialized Agency of the United Nations, founded in 1944 through the Convention on International Civil Aviation signed on December 7, 1944. ICAO is charged with the responsibility to establish international standards, recommended practices and procedures covering all aspects of international civil aviation operations. These include technical, economic and legal issues.

In order to fulfill its international obligations, ICAO has a staff of international representatives who can provide valuable expertise in all aspects, including the identification of political sensitivities that need to be acknowledged in defining strategies for change. The effectiveness of the organization further depends on the expertise and input of other organizations. These include international regulatory authorities, safety boards, the International Air Transport Association (IATA), an industry-lobbying group representing over 250 airlines, the International Federation of Airline Pilot's Association (IFALPA) and the International Transport Workers' Federation (ITF) as some of the main examples, to accomplish their mandate. From the perspective of the air traveler, these organizations have their own vested interests at heart and do not directly address the consumer's concerns. The author's own research (1995) with randomly selected users of Canadian air carrier revealed that only a slim minority was familiar with the role of Transport Canada and still fewer with any other aviation organization.

Since it publicized its strategic plan in 1997, ICAO commits itself to a more active role in ensuring signatory countries to the Convention apply the safety and security Standards and Recommended Practices. The current focus is on Safety Oversight and Unlawful Interference programs, the latter also addressing the issue of disruptive passengers. As such ICAO has a major role to play in the tightening of the international legal safety net.

An Example of a National Aviation System

The following diagram illustrates the relative relationships between the Canadian air carriers and the various national and international organizations and associations that are part of the aviation system.

The list includes the key players such as: the regulatory authority Transport Canada (TC), the Transportation Safety Board of Canada (TSB), the Canadian Transportation Agency (CTA), the Airline Pilots Association International (ALPA), the Air Canada Pilots Association (ACPA), the Airline Division of the Canadian Union of Public Employees (CUPE), the Air Transport Association of Canada (ATAC), and the Air Passenger Safety Group (APSG), an arm of Transport 2000 Canada.

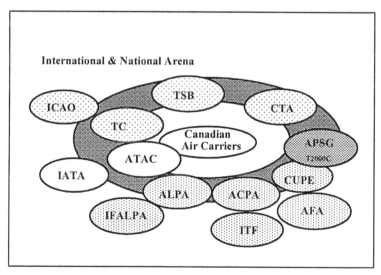

Figure 5.1   Canadian Aviation System

The inclusion of international organizations has been limited in this figure to a total of five. The International Civil Aviation Organization (ICAO), the International Air Transport Association (IATA), the International Federation of Airline Pilots' Association (IFALPA), the Association of Flight Attendants (AFA), and the International Transport Workers' Federation (ITF) are not the only organizations outside of Canada with a distinct influence on its aviation system. Transport Canada has a long-standing relationship with the Federal Aviation Administration of the United States, the same as the Transportation

Safety Board maintains close relations with the National Transportation Board to name a few. The web of relations is far reaching across the boarders with the view to share information and protect interests.

In Canada, the operative word is partnership, partnership between the regulatory authorities, industry and labour. Its dynamics were relatively stable for many years. Until the merger of Air Canada and Canadian International Airlines, only these two operators with their regional connector carriers in tow were the dominant players. The merger further unbalances the limited lobbying power of the few and less experienced independent carriers via ATAC, which will now be dominated by merger issues involving Air Canada for some time to come. The growing public discontent with the perceived deterioration of service from the merged Air Canada places pressure on the government for tighter regulations, including increased competition. It is unlikely that this can occur from Canadian carriers unless foreign carriers are granted greater freedom to access currently protected routes.

Air Canada's CEO, Robert Milton with a leadership style that is well suited to the needs of the organization and its future aspirations in the global arena has firmly signaled a new era of relationship with the government, regulators and the business community at large. His strong formation in the American aviation industry prior to joining Air Canada will shift the traditional way of easy cooperation to one of greater distance and more conditional cooperation.

Consumer Protection

Statistical consumer information on the quality of airline services is not commonly available. In most cases, airlines consider this information as proprietary and do not publicize their internal complaints statistics. Legislation is one way to guarantee consumers access to this information. The Department of Transport in the United States issues monthly statistics on scheduled airlines, addressing the number of cancelled flights, passengers being bumped because of overbooking, lost baggage, and consumer complaints. This is not the case in Canada although the introduction of Bill-C26, currently being reviewed by the House of Commons addresses this issue to some degree.

Bill C26 is aimed at preventing Air Canada, now in a monopoly position, and applicable to other scheduled and charter air operators, to abuse their positions. In line with the goals of Bill-C26, the public will be able to direct its complaints, in writing only, to an Air Travel Complaints Commissioner who in turn will provide the Transport Minister with statistics every six months.

Critics feel that these efforts are insufficient and do not adequately protect the consumer. Clearly, without the benefit of timely performance data on key service aspects, the consumer does not have a basis for informed decision-making other than what is sporadically reported by other individuals or the media. Bi-annual reporting has no value for the consumer other than acting as a placebo. Under these types of conditions, airlines will continue to enjoy greater protection than the consumer. They will be able to conveniently explain periods of increased service glitches long after the fact and minimize the impact on the consumer. The inconvenienced or mistreated air traveler is still left to fend for himself by directing his complaints to the airline in an effort to obtain compensation that varies from airline to airline. Consumer advocacy has not been a great force in Canada, in part because of the more tolerant nature of Canadians compared to Americans, the small population, significantly lesser air traffic, and the established mechanisms through regulatory channels.

The Air Canada merger catalyzes this change to some degree. It awakens consumer memories of previous discontent while the effects of merger activities reveal themselves in the day-to-day operation. They sharpen perceptions of current experiences and unite the consumers' cry for fair and better service. While the bigger-is-better syndrome heartens stakeholders, the consumers rally their concerns with a focus on schedule convenience and frequency of flights, to customer service issues. In addition, Canadian air travelers' summer holiday plans in 2000 are jeopardized by the Air Canada flight crews' contract dispute. The Canadian Transportation Agency (CTA) received 257 written reports in the first six months of 2000, a 59 percent increase of passenger complaints over 1999 for the same time period. CTA introduced a hotline at the beginning of July 2000 registering 223 calls over the first eight days of operation.[11] It will be interesting to monitor Canadian consumer discontent, and measure what effects these developments have on regulatory initiatives, consumer advocacy and airline compensation policies in future.

A Complex System to Serve the Public Intertest

The causes for passenger frustration are related to the chain of events that occur within the aviation operation including contract disputes, and the lack of an aviation infrastructure capable of handling current capacity demands. Ultimately they affect what is taken place in the passenger cabin. The passenger may not be aware of these events until after boarding the aircraft nor does he necessarily care about the larger picture. What is important is the immediate experience as to how the flight and cabin crew manage this situation. Passengers get nervous and anxious without timely and effective information on what causes a delay in leaving the gate or in take-off clearance. The fact that the crew may have great difficulties in obtaining useful information is of secondary concern to the passenger.

Increased automation results in greater risks to system failures creating at times chaos at airports around the world. One such incident occurred at London's Heathrow airport over a weekend in June 2000. It took over two days before schedules were back to normal.[12] Another mishap forced

Northwest Airlines to cancel 130 of its 1,700 daily flights and delay hundreds more around the world when its computer network failed after a utility crew accidentally severed a crucial link to its internal data network.[13]

In an effort to increase runway safety, NASA continues to test its system designed to more accurately predict wake turbulence generated by aircrafts at Dallas/Forth Worth International Airport.[14] Called Aircraft Vortex Spacing System (AVOSS), it would allow air traffic controllers to more efficiently space aircraft during approach. The selling points are safety and 15 percent increase in runway capacity.

Air Traffic Control

Air traffic control (ATC) is an interactive system involving both aircraft and ground-based ATC. Its success relies on the operator continually adjusting to an evolving technical system at the operational level while maintaining or improving his/her performance. The demands on the operator are high since complex automation continually advances flight deck systems and ground-based management system that are not always fully integrated.

Air traffic control is responsible to monitor flight movements according to strict rules and regulations. Controllers give direction to flight crews in setting their course for approach, cruise and landing, detect hazardous flight path conditions, respond and correct altitude deviations, and make split second decisions to safeguard aircraft movements in all types of weather conditions. The performance of ATC has a direct bearing on airlines' on-time performance and passengers' satisfaction or resulting stress level.

The controllers work in a highly technical environment, demanding a high degree of cognitive skills, efficient and quick decision-making. The work is stressful, especially at airports with high traffic volume. The system is error intolerant with many safeguards in place to ensure the safe movement of airplanes. Notwithstanding, technical system failures, inadequate equipment, miscommunication, misjudgment or inattention could easily cause an accident. In one of many examples of equipment inadequacies, The National Air Traffic Controllers Association in the U.S. called on Congress to release funds for upgrading the 30-year old software.[15]

Apart from technical resources available to ATC, human factors play a major role in the smooth functioning of the system and aviation safety. Air controllers in France have identified a 50 percent rise in near-midair collision in 1999. The union cited an increased workload as the major cause for these errors.[16] Similarly, air traffic controllers reportedly caused massive delays or cancellations at O'Hare International Airport in July 2000.[17] The situation might not be that simple; the FAA cited strong upper-altitude winds as the main reason.

In Canada, the privatization of Air Traffic Control is also fraught with growing concerns. Although not an issue receiving public attention, worker grievances are mounting and similar to those expressed in other countries. Some of the common concerns are:

- Shift schedules, especially 9 or 10 hour midnight shifts;
- Single controller during midnight shift;
- Staff shortages;
- Controller fatigue;
- Time off in lieu of overtime;
- Training;
- Inability to schedule mandatory Occupational Safety and Health meetings;
- Ineffective Critical Incident Stress Program;
- Poor management – labour relations;
- Division of airspace due to centralization;
- Maintenance of equipment.

Human Factors' research is expanding in air traffic control with the aim to predict the evolution of human performance when interfacing with the high-pressure technical environment and increasing traffic volumes. The situation is serious enough to make all efforts to reduce air traffic control errors. Assessments models such as the SHEL model (Edwards, 1972) and the later modified SHELL model (Hawkins, 1984) identify the interactive elements of software, hardware, environment and 'liveware', the controller.

Despite those efforts to understand the nature of human error, and develop strategies to reduce these incidents, the circumstances under which they occur have not changed. They are likely to worsen. Production and performance pressures increase with traffic volume lack of experience and during the first 15 minutes of shift change (Isaac & Ruitenberg, 1999).

The effects of long-term organizational disputes between management and labour in ATC organizations, especially noticeable in France in the summer of 2000, are delaying flights and inconveniencing thousands of passengers. The organizational climate and culture in which these disputes occur cannot be separated from their impact on the air traveler at the receiving end of these upheavals. Because of the system impact, air travelers worldwide are experiencing first hand the effects. Organizations embroiled in these disputes are far removed from the public and the airline workers facing the consequences. For the passenger, the shifting power base of the multiple components affecting their travel experience may never be clear without concise communication.

The question of management responsibility versus cost control and maintaining safe-working conditions is no longer simply an internal matter. A large percentage of ATC management is formed by their operational background, thus lacking the skills to apply the type of people skills needed in today's complex workplace.

Airport Design

Speed is of essence: getting to and from the airport, moving through, and boarding or deplaning the aircraft. The air travelers' stress is closely linked to the ease or lack of accessibility from the home, road infra structure, parking facilities or public transportation to the airport.

The airport design itself either facilitates or hampers the air travelers' movement from the check-in area to security and the gate. Line-ups impose stress. Long walks through the airport facility to the boarding gate are also undesirable from the travelers' point of view. Especially the elderly, parents with small children, the disabled and the less fit traveler are often challenged by the distance they need to cover and the lack of accessibility aids.

Signage, noise level including frequent announcements, distractions of shops, and crowded conditions are contributing to the air traveler's sensory distraction.

Adequate gate and bridge facilities, especially at older airports coping with increased air traffic, can cause delays in aircraft movement and passenger boarding and deplaning. At government owned airports, the effects of government funding or lack thereof are visible in the general condition and development of facilities including runways that can handle the number and types of aircraft scheduled. Privately owned airports, such as in Canada, operating costs and expansion requirements have resulted in some airports levying additional airport taxes on the air traveler. Forcing them yet again to line-up for making these payments adds frustration. Intensifying these pressure points are often surly, condescending, arrogant responses, or soft-toned courtesies carrying their own aggression.

In the U.S., consumer research indicates that airport improvements are very high on the public's wish list. Nearly 1 out of 4 responses expressed the need for better facilities. Subsequently, the National Air Transportation Association (NATA) has launched its initiative in identifying 100 of the nation's most needed airports as targets for funding. Criteria for analysis include growth, use and local impact balanced against airport capacity.[18]

Apron Management

Apron management deals with the support functions necessary in dispatching departing and arriving flights. Adequate gate facilities and ramp space are essential for the time sensitive choreography of ground support to take place. Any congestion, improperly placed equipment, lack of experience by operating personnel or error can seriously affect the on-time departure and arrival experience for the air traveler. The availability of an adequate number of aircraft power units (APU) is another example that if unavailable, the aircraft cabin is subjected to the weather, be it freezing or swelteringly hot.

On a recent flight, an associate of DAHLBERG & ASSOCIATES, together with her fellow travelers, was held captive on board for twenty minutes upon arrival, simply because the ramp staff was inexperienced and could not get the bridge adjusted to the aircraft boarding door. Passengers seated in window seats craned their necks to see what was taking place, after the captain made an announcement. During the long delay, passengers started to gesticulate to the person operating the bridge, shaking their heads, laughing and finally resigning themselves back into their seats, except for a few. Comments accompanying their actions were clear. They were not impressed; they readily blamed the airline for this poor performance.

Airplanes

The interface of ground support functions and effective communication with the flight crew benefits or detracts from the air traveler's experience of a flight. Problems related to aircraft maintenance, cargo and baggage loading are some of these examples. At times, serious implications for the safe operation of the aircraft result when related to improper hazardous goods handling and weight and balance issues.

At best, the problems are identified in a timely manner on the ground, however forecasting the time required to resolving these problems is not easy, especially when original estimates turn into a creeping delay. This leads to potentially tense situations on-board, not only for the time the problems are being dealt with on the ground but carrying over in-flight. Both the crew and passengers are challenged to cope with a range of negative responses to such situations.

Weather sensitivity of the operation and the ability to manage these conditions to the satisfaction of their customers are not always under the control of the airline. Safety regulations concerning electrical storms shut down all ground functions involved in getting the airplane ready for departure or arrival. Air travelers who are caught on board without the ability to deplane or depart on time demand to be kept informed and need a reassuring cabin crew to monitor passenger responses and act caringly with the means available to them.

Illustrating the impact of weather is the report by the FAA of 153 occasions in July 2000 of thunderstorms affecting operations at major U.S. airports, up 99 instances from July 1999. Cooperation between government and the airlines resulting in improved sharing of weather data, rerouting planes and better decision making for earlier cancellations of flights will improve the overall planning process.[19] The impact of 44,401 delays during the same month still cannot be blamed on weather alone although there is a ripple effect throughout the operation.

Crew shortage as a result of poor weather conditions and delays is linked to contractual limitations for monthly maximum flying time. Again, the air traveler is inconvenienced facing flight cancellations and delays. The airlines have little choice but to take these measures since staffing parameters for overall coverage of schedules are already including a buffer for a number of contingencies.

The Inside Story

On-board relationships between crews and air travelers are never simple. As the woes of United Airlines in the summer of 2000 reveal, a labour dispute with its flight crews and mechanics, combined with weather related delays and equipment failure, have resulted in the worst on-time performance of all major U.S. carriers, and an unprecedented number of approximately 2000 domestic flight cancellations in August 2000. The public does not want to be at the mercy of labour disputes, and blames corporate management: the 'current climate feels more like contempt than competition.'[20] As the official carrier for the Olympics in Sydney, athletes and team organizers are also very much concerned despite assurances by United Airlines that these flights will not be impeded.[21] Compounding anxieties will taint passenger perception of the quality of service and add to the general tension onboard.

For the crew, a delay can mean conflict with their duty day as stipulated in their contract. At worst, the crew is in a position to advise their company that a full or partial replacement crew is required. This leaves the passengers further exasperated being caught in the vortex of a chain reaction that they have no control over. The replacement crew faces additional challenges under these circumstances. At best, they obtain detailed information from the outgoing crew on special attention passengers, general social atmosphere in the cabin, provisioning status, and what information has been given so far. The exceptional cabin crew team immediately plans for damage control in dealing with the already aggrieved passengers, although all too often, the desired communication does not occur.

Some crews are dealing with these situations with professionalism, well aware that communication with the passengers is key to maintaining social peace and goodwill. Their attitudes, experience, training combined with their on-board leadership and the labour relations climate are very much linked to the successful conclusion of irregular operations.

On-board service initiated on the ground during a lengthy delay may be well received although there are other concerns to be addressed. Safety regulations, the airlines policies regarding uplifting supplies and the caterers' ability to replenish cabin crew requests in a timely manner, are some of the limiting factors. The cabin crew may well be willing to provide additional service, however their authority to secure the necessary supplies or a station's ability to comply with these requests is another matter.

The cost of extra services is a significant concern for the airlines. Catering contracts usually include additional fees for special provisioning runs and supplies. Cabin crews are therefore subject to company scrutiny and have to be able to justify their requests. The Station Operations Control Centres usually act as intermediaries when advised of a special request. Cabin crews are to follow a clearly defined communication path in identifying specific requests in a timely manner. Even so, the Duty Manager will consider all aspects of discrepancies affecting the operation at that time and may decide in favour of on-board provisioning shortages in order to maintain on-time departures to protect down-line connections as some of the overall operational decision criteria.

Knowledge of passenger demographics, cultural background, class of service, and purpose of their trip are elements the cabin crew must take into consideration in determining their course of action. Chapter 4 addressed these elements in greater detail.

In airline merger situations, customer loyalties and attitudes towards the airline in a take-over position find new expressions in the way they respond to the cabin crew. Customers play out national East-West divisions in Canada with unsolicited verbal aggression toward the cabin crew. This is not an easy way to start a flight.

Politics

National political unrest also has an impact on airline operations and its customers. Flight cancellations such as experienced in Israel in October 2000 are one of the many examples in recent history, and can have a major impact on international airlines and their passengers.

Marketing – the Driving Force

The Hard Side of Business

The aviation industry is driven by facts and figures. Corporate productivity measurement criteria are evidence of this. The record profits achieved by American air carriers in 1998 and 1999 of US$9.3 billion and US$8.5 billion respectively are a success story for corporate accomplishment and celebrated with lavish bonuses for its industry leaders. The focus on shareholder value with pressures on profits did not necessarily result in improvements in efficiency, customer service, and managerial effectiveness. At no time was the public outcry more intense than in 1999 and continuing in 2000 with consumer advocacy harnessing politicians to force airlines to provide major improvements to their service.

Delays are leading complaints. In the U.S., Department of Transport statistics for the first quarter in 2000 reveal Fridays are consistently the worst day of the week for late arriving flights while Wednesdays are the best.[22]

Delta and United Airlines recognize passenger frustration during delays and look for technology to divert potential hostility. Delta Airlines is already offering wireless Internet service at Vancouver's International Airport and plans to expand this service to its major hubs in the U.S.

United Airlines suffering its worst on-time performance in August 2000 with an on-time performance of only 42.7 percent also plans to introduce this service in 2001. At a monthly charge of US$40.00 to US$60.00, frequent flyers are able to purchase unlimited access to the Internet. Benefiting he most immediately are the major laptop makers manufacturing computers that can handle wireless connections and the radio waves.

Attention to corporate profits and sheer greed overshadow the necessity to respond to the unique and often ill understood air traveler needs in the overall service design. I addressed these needs in greater detail in Chapter 2. With pressures for shorter turn-around times, keeping slot times, increasing congestion at major airports around the world, the responsibility for softer side of business is entirely left to cabin crews. On-time performance is a key in the operational world. Galley shortage reports and deferred cabin logbook entries are clear measures of the hidden impact of on-time performance, resource allocation and supplier relations. Few airlines manage to reconcile productivity and service objectives successfully.

Although the entrepreneurial spirit of private and publicly owned businesses has been credited to be the key engine for a market driven economy, a lasting transformation involves more than leadership, trendy programs and band aid solutions. In Western airline organizations corporate leaders come and go regularly while flight crew and cabin crews are part of the more permanent fabric of the organization. They form a sub-culture, apart from the rest of the organization with their own values and norms. This is an important aspect to consider when new management attempts to influence the crew culture.

The Challenge of Delivery

Airline organizations have been in a state of transition since deregulation. The immediate reaction to the power shift within the airlines organizations was concern by the Flight Operations and the In-flight Services departments, especially in respect to the impact that was felt of service improvements on the state of airline safety. These concerns are universal to this day, waiting to be resolved through a synergistic approach to safety and service.

Marketing translates 'needs' mostly into tangible products. In the passenger cabin, it is seat configuration and seat comfort, including all fixed systems and décor designed to be pleasing and efficient. It is the stage set, permanent to a great degree and serving as a backdrop to on-board catering and the delivery of the goods. The product is transient, its quality highly dependent on a wide range of suppliers and their mutual relationship. The customer sits in judgment while experiencing its delivery. Products of this kind appeal to our senses in different forms, real and consumable or in advertising. They are a lot of fun to play with in the planning stage. To deliver these products in the spirit they are conceived is a different challenge.

Technological advancement in the passenger cabin have led to much improved customer entertainment options and effects the relationship with cabin crews by creating different expectations. The cabin crew plays an important role in trouble shooting should the systems fail. Consistent delivery of electronic entertainment lies in the area of system reliability and maintenance. Failure of these services produces passenger annoyance and frustration on part of the cabin crew since not only do the have to attend to the mechanical failure at hand but are now also faced with a large number of annoyed passengers.

As the delivery of tangible products such as meals and beverages has become easier based on improvements in galley design and equipment on newer generation aircrafts, the complexity of some of the services based purely on technology, have added different demands on the cabin crew. Passengers view cabin crews and their ability to deliver all goods as part of the overall corporate product, and so do many airlines. Increased technical knowledge of entertainment systems and their operation, finesse and timing are a basic challenge in a time driven environment. The greatest challenge lies in linking product delivery to the cabin crews' motivation to keep their routine tasks fresh, sensitive to customer needs and personalized.

Policies and Procedures

Company policies and procedures can have a fundamental effect on deterring or enabling abuse and violence in the passenger cabin. Most important is a no tolerance policy, backed up by a corporate safety net for front line personnel. This safety net must draw its support from a wide range of organizational entities, including marketing, security, airport personnel, human resources, flight operations, in-flight services, labor associations, health and safety, and training.

North American and European carriers who have instituted such policies have achieved a significant change in employee attitudes towards their management. Some North American carriers have taken action to deter extreme passenger misconduct using temporary travel ban and, loss of flyers' points as a means of deterrent. This visible and published response gives much needed support to the employees who are faced with abuse and violence in their workplace. Due process required to safeguard against unfair punishment.

Changes in regulatory policies and procedures can also cause increased passenger tension. This was evidenced by the high number of incidents relative to non-smoking infractions and resulting passenger misconduct following the introduction in North America and on international flights. Asian carriers who only recently have introduced a no-smoking policy report similar trends.

South African, now featuring smoking restrictions at airports and in flight, account for an estimated 50 percent of violent on-board incidents, a concern raised by ALPA-SA. The pilot association looks for harsher sentencing of disruptive passengers.

Service Product Design and Procedures

One of the more contentious issues is the role of alcohol in the design of the on-board service product. Paid bar versus complimentary bar policies are driven by competitive considerations without a base of social or corporate responsibility.

Free bar service on charter flights and long-haul international flights are a recipe for creating conflict on-board. The cabin crew is charged with the regulatory responsibility to assess and discontinue serving alcohol to a person showing signs of intoxication. Consistency in carrying out this responsibility is a different matter.

The consumer has a set of expectations linked to what the marketing promises of the airline. If it is free alcoholic beverages, entitlement comes into play, especially with those individuals who abuse alcohol. Statistics clearly connect sports and vacation groups, transient workers and certain nationals with a greater disposition to consuming larger amounts of alcohol than other airline travelers.

Canadian cabin crews operating international flights from Europe to North America are familiar with the heavy drinking patterns of ski groups for example who come to have a good time. After boarding their flight the first thing they request is a beer and a cognac. At the end of these flights, the on-board bar is almost depleted. The drinking goes on for the duration of the flight, loosening tongue and voice, the group erupting in song without consideration for the rest of the passengers. Their loutish behaviour is part of a group culture based on macho values and devoid of civility. This causes conflict with the cabin crew when they attempt to curtail both consumption and merriment.

Video Gambling – a Threat to On-board Safety?

In the search for additional sources of revenue, video gambling on-board is one of the options considered by some airlines. In March of 1996, the Department of Transportation submitted a report to the Congress, 'Video Gambling in Foreign Air Transportation.' In this 57-page report safety effects, competitive consequences and bilateral issues and legal framework are at the center of analysis.

The report examined the competitive consequences of permitting only foreign air carriers to offer video gambling. With an anticipated shift of 4 percent in international traffic in the Atlantic and Pacific regions, an annual loss of over US$ 490 million to U.S. airlines was estimated. The cost of installing such systems at that time was estimated at US$401 million, a one time capital expenditure, while the additional fuel cost due to the weight of these systems was estimated at US$43 million annually. It was further estimated that 18 percent of passengers would use an on-board video gambling feature. It was further estimated that gambling would generate US$480 million in annual revenues for foreign airlines. Clearly economics are a major consideration in proposing changes to the current law prohibiting gambling on all commercial flights within the airspace of the United States and in fifth-freedom markets.

Chapter 3 of the report deals specifically with 'Safety Effects of Electronic Gambling Devices On Board Commercial Aircraft.' The section of interest to us deals with the issue of behavioural risks. The Association of Flight Attendants has already registered their concerns that are mainly in the areas of handling problem gamblers, handling money for gamblers, access to gambling by minors, and increased workload with the potential of interfering with their safety related duties.

While at the time of the report no experience with video gambling related problems on an aircraft had been gathered, studies reveal a grim picture of compulsive gamblers. The American Psychiatric Association first recognized compulsive or pathological gambling as a mental disorder in 1980.

The extent of gambling and its impact on American society is becoming more acute and concerns are spreading as reported in 'Pathological Gambling: A Critical Review' published by the National Academy Press, Washington, DC 1999. Reportedly, over 80 percent of American adults gamble at some time during their lives. In 1997, the year for which the latest economic figures are available, gambling accounted for more than US$551 billion in wagers. The intense competition between states for moneys generated, have liberalized legislation in all but three states. The Commission on the Review of the National Policy Toward Gambling estimated in 1975 that less than 1 percent of the population were 'probable compulsive' gamblers. New technology is adding to the spreading problem. The current estimate stands at 1.5 percent of adults falling into this category with 1.1 million young adults between the ages of 12 and 18 estimated to indulge in gambling on an annual basis.

Studies also confirm that pathological or compulsive gambling occurs with other disorders such as substance abuse and antisocial personality disorders. These findings are not encouraging for cabin crews having to face the potentially disruptive behaviour by such individuals on their flights. Regulators are aware of these facts however the economic benefits to the airlines weigh heavily on concerns for engineered social unrest in the passenger cabin.

Opposing Values

Opposing values result in different cultures within an organization slowing progress in closing the rift between safety and service. The broader context of the aviation system shows similar trends. In the international arena Contracting States are not accountable for planning and implementing improvements. ICAO operates on the concept of consensus building. The process is difficult and fraught with bureaucratic and political considerations. The lack of commonly shared values and resources is hampering the advancement of airline safety globally. This is a predictable consequence, as divergent values in any organization must have serious consequences on their effectiveness and ability to survive.

For more than two decades, the traditional hierarchical organizational structure of all types of industry has been under severe attack. Management experts see the rigid structure as ineffective in responding to the environment of fast-paced, increasing global competition and exploding information technology. New concepts have evolved viewing organizations as ecosystems, dependent on a global system. Now, organizational success is largely determined by relationships and interaction with the external environment.

The traditional pyramid structure is seen as no longer meeting the demands of rapid change, growing instability and unprecedented leaps in technology. Transitional concepts such as the 'inverted pyramid' are evidence in the struggle to create a new vision of a new reality. However no matter what new concepts and terminology is introduced, at the heart of the ongoing debate remains one question: How does the organization deal with the relationship between managerial control and accountability relative to preventing social unrest on-board?

The aviation community treats its customers showing signs of distress in a predominantly adversarial manner. The emphasis is on establishing the rights of the airline worker as an enforcer of rules and regulations with diminished empathy for the individual that appears unable to readily conform.

The consistent demand by crew labour associations for harsher punishment illustrates the firmly rooted beliefs of their members that only the legal system will effect social change.

Methods used to analyzing on-board incidents further reveal the adversarial nature of the relationship between the airlines and their customers. A review of a cross-section of investigative reports shows that it is predominantly a one-dimensional analysis with a clear focus on the offending customer without the benefit of examining other factors such as company policies and procedures, the physical environment in which the conflict occurs and other related pre-conditions impacting on the interaction between the air traveler and the airline worker.

Cases where the mental stability or illness of the passenger is a major contributing factor require a set of skills by the airline worker outside the common approaches recommended for conflict resolution. As Renee Sheffer recounts, her previous training as a psychiatric nurse enabled her to diagnose the situation and apply her skills successfully to a point. The well-meaning assistance of macho-type passengers unfortunately exasperated the situation that she may have been able to conclude without suffering major injuries and trauma, as was the case.

In the case of the male passenger who broke into the flight deck, lunging for the control and terrorizing crew and passengers alike on an Alaska Airlines flight in March 2000, medical experts agreed that a rare form of encephalitis[23] caused his violent outburst. In conclusion, the man was not responsible for his actions based on this diagnosis[24], however has been ordered to undergo psychiatric testing.[25]

The variety of causes triggering unpredictable violent behaviour on board is growing as individual cases are being investigated. Realizing how little airlines and the public in effect know, should lead to new ways of dealing with these situations. Reactionary strategies based purely on fear and ignorance block the development of a broader, more compassionate approach.

The current emphasis on preventive measures is overwhelmingly aimed at training employees in their rights and responsibilities, including the interface with enforcement agencies. This trend is consistent with a reported loss of 'dignity and respect for the authority' of airline workers, especially flight and cabin crews. We have seen a significant shift away from the 'Nightingale of the Skies' role of cabin crews to one of guardians of air regulations and defenders of their rights.

Although the trend to educate workers and the public on their respective rights and responsibilities is one aspect of a strategy to curb passenger misconduct, the important aspect of empathy with the plight of the customer, especially those suffering from illnesses, has not been adequately addressed at any level.

Leaders in the aviation industry must concern themselves with issues of organizational commitment to social peace on board. The training programs are important, however they need to address Human Factors, medical and psychiatric issues that have become more prevalent in today's society. Social peace on board requires a more physically tolerable cabin environment, especially in economy class.

Policies and procedures, including those concerning the serving of alcohol, need to be addressed more effectively to allow cabin crew to manage potential incidents. Marketing departments have tried to remove themselves from assuming their responsibility by rationalizing their decisions on the actions of the competition. A common response to inquiries is: 'We would certainly consider it (paid bar) if our competition would do the same.'

Applying Reason's Error Model

Reason's model of latent and active failures can be applied with certain modifications as indicated in *Figure 5.2* to organizational factors enabling rather than reducing opportunities for passenger misconduct. Although the model originally aims at providing understanding of failures leading to industrial or aircraft accidents, it also offers parallels to the management of preventable on-board passenger incidents.

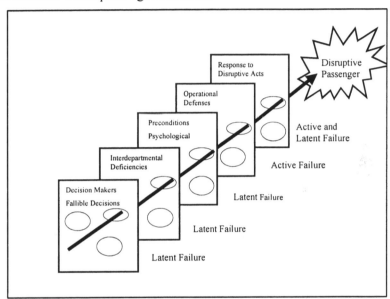

Figure 5.2   Reason's Modified Error Model

The major difference lies in the basic premise that fallible decisions by corporate decision makers enable the occurrence of preventable passenger misconduct. They constitute potential stressors for both the airline worker and the passenger.

The primary origins of passenger misconduct however are highly complex, and include passengers' psychological precursors. Factors of the passenger him/herself, unknown to the airline worker are being discussed in greater detail in Chapter 6.

Latent failures as shown in this model refer to three aspects of the organization:

- The first identifies corporate leadership as evidenced in goal and policy setting, e.g. failing to establish a zero tolerance policy on passenger misconduct.
- The second identifies failures to coordinate interdepartmental policies

and procedures in support of a corporate zero tolerance policy, including marketing strategies and product design.

- The third identifies preconditions such as reliable service products, well-trained and skilled employees. Psychological precursors include motivated workers with an appropriate set of attitudes, work schedules, work environment, and cultural norms conflicting with the goal to minimize social risks to both the safety of employees and passengers.
- The fourth identifies operational defenses. In the case of airline operations this addresses pre-board screening, e.g. denying boarding of intoxicated and visibly disturbed passengers. It also deals with the management of irregular operation, delays, and over-booking situations that are known stressors for the air traveler.
- The fifth refers to the performance of the airline worker, the delivery of the promised service, and dealing with a difficult passenger situation.

Latent failures, described as loopholes in the system, are human failures that only become apparent when they become activated by circumstances (Reason 1990). The airline organizations rely on their customer contact workers and cabin crews to recover service failures that to a large degree have been deliberately planned as part of the overall risks associated with doing business. Some examples are over-booking, inadequate training, lack of adequate data gathering systems, inadequate customer service recovery programs, choice of cabin configurations.

Active failures, described as slips, lapses, mistakes and procedural violations occur just prior to and/or during the interaction with the customer.

Latent failures occur at the top of the organization and are the result of incompatible corporate goals where safety and service are designed to conflict. They are then played out in interdepartmental deficiencies such as an inability to address these conflicts effectively, poor communication, and rivalry for corporate resources. Monies spent on behalf of marketing enhancements by far outweigh monies spent on cabin crew training. In light of the significant impact these decisions have on cabin crew performance, it may well be timely to allot some of these funds for more adequate training of these workers so that they may deliver the total product with the necessary skills.

Psychological and attitudinal precursors include conflict with management and colleagues from other departments. A ground agent insisting to board an intoxicated passenger or failing to take action with excessive carry-on baggage are two of many examples; excessive workload due to an overload of service during critical operational times is another. A culture that promotes the principle 'the customer is number one', or 'the customer is always right' is equally contributing to avoidable passenger misconduct.

Operational defenses such as training, equipment, experience, attitudes, skills and a supportive organization may be inadequate and contribute to both active and latent failures when confronted with passenger misconduct. An example of this would be the absence of restraining devices on-board, inadequate cockpit door design, and lack of training in restraint procedures and backup. Dealing with irregular situations without the benefit of clearly defined procedures and communication systems, is another.

The degree of physical protection afforded to the crew, including the availability of effective restraining devices for use by the flight and cabin crew can make a difference in rendering the out-of-control passenger harmless. A number of international airline operators have already instituted these devices while others will soon carry handcuffs for use in restraining passengers who become violent.[26]

Concerned about the degree of protection afforded to flight crews operating DC-9s and MD-80's, some airlines in the United States are examining the possibility of reinforced cockpit doors. The standard bi-fold cockpit door is structurally less sound as evidenced by an incident in March of 2000, when a 250-pound man broke through the door of an Alaska Airline flight, attacked a co-pilot, lunged for the controls and shouted: 'I'm going to kill you.'[27]

Evidence of such latent and active failures clearly need to be rectified with a clear focus on protecting crews, workers interacting with the public on the ground, and air travelers at large. As countless incidents reveal, the main defense system is based on the individual cabin crew member or that of their colleagues, their awareness, their instinctive reaction, their self-developed skills in handling conflict.

Reason's failure concept, although developed to assist in the analysis of accidents, has a place for application in the management of disruptive acts. Airlines could profit from directing their efforts at closing the loopholes of latent failures within the organization rather than at recovering once an incident has occurred.

Conclusions

Serving the public interest in civil aviation is a complex undertaking. It requires governments, legislators, regulatory authorities, aviation organizations, industry, labour, and aircraft manufacturers to work together in new and more informed ways. The air traveler as an essentially human factors element deserves greater recognition in the overall design and operation of the aviation system.

Without an understanding of historical developments in aviation and their impact on current efforts towards a socially safe environment on-board, progressive steps may be poorly integrated. Education of both the airline workers and the public should be part of the airlines' responsibility to this end. The media is already making a contribution by a noted improvement in their reporting. Operators of the aviation system not facing the customer in their work need to be sensitized to realize their responsibility in contributing to undue passenger stress and frustration.

Marketing, with the difficult role to achieve competitive advantage, is too intent of using tangible products to cater to the premium customer while failing to examine the social impact in the passenger cabin or the risks of additional service enhancements for the sole purpose of on-board entertainment.

Airline monopolies and their leaders, just stopping long enough to contemplate, may find themselves realizing what an organization is without the good will of their customers and their workers.

Airline organizations must act on the opportunity to realign their safety and service goals. If it is not for reasons of concern aimed at their customers or employees, it should be simply smart. Airlines who accomplish this will no doubt be more successful in a competitive environment which is facing an increasingly sophisticated customer base and will also benefit from insurance underwriters willing to reward such corporate decisions with substantial reductions in liability insurance. All around, no matter what the final motivation, smart organizations will benefit in many ways.

Notes

[1] The Tenerife runway collision in 1977, the Three Mile Island accident in 1979, Chernobyl and Challenger disasters in 1986 are some of the large-scale events that jolted attention to human error research and examining underlying organizational issues contributing to these tragedies (Reason, 1990).

[2] *Financial Post*, 24 June 2000.

[3] *Reuters*, 24 May 2000.

[4] *Avflash*, 21 February 2000, vol. 6, 08a.

[5] *AVflash*, 10 February 2000, vol 6, 06b.

[6] *Avflash*, 9 March 2000, vol. 6, 10b.

[7] *Calgary Herald*, 6 June 2000.

[8] *Reuters Ltd.*, 31 July 2000.

[9] *Avflash*, 10 July, 2000, vol.6, 28a.

[10] *Calgary Herald*, 27 July 2000.

[11] *Calgary Herald*, 19 July 2000.

[12] *Calgary Herald*, 19 June 2000.

[13] *Airjet Airline World News*, 23 March 2000.

[14] *ABCNEWS*, 17 July 2000.

[15] *Avflash*, 3 February 2000, vol. 6, 05b.

[16] *Avflash*, 3 February 2000, vol 6., 05b.

[17] *The Associated Press*, 18 July 2000.

[18] *Avflash*, Vol. 6, 32a.

[19] *The Seattle Times Company*, 11 August 2000.

[20] *Airlinebiz*, 30 July 2000.

[21] *Associated Press*, 13 August 2000.

[22] *USA TODAY*, 30 June 2000.

[23] The National Institute of Neurological Disorders and Stroke in the U.S. describes this condition on their website as follows: 'Encephalitis is an inflammation of the brain. There

are many types of encephalitis, most of which are caused by viral infection. Symptoms include sudden fever, headache, vomiting, photophobia (abnormal sensitivity to light), stiff neck and back, confusion, drowsiness, clumsiness, unsteady gait, and irritability.'

[24] *Associated Press,* 20 October 2000.
[25] *Associated Press,* 26 October 2000.
[26] *Airlinebiz,* 4 April 2000.
[27] *Airlinebiz,* 23 March 2000.

6 Passenger Risk Management

Introduction

The pressures facing the aviation community in finding solutions to the apparent mounting incidents of passenger misconduct have led to a broad range of initiatives at a number of levels. They range from new laws in some countries such as the United Kingdom and Canada, to changes in cabin crew training programs, from passenger awareness campaigns to increased awareness in the legal profession and airport enforcement units. Improved international cooperation is becoming evident, although slowly.

Organizations such as ICAO, IATA and the International Transport Workers' Federation (ITF) have a major role in developing an action plan that will lead to a more effective legally enforceable international safety net.

Diverse powers of jurisdiction in each country facilitate or hinder charging the individual involved in disruptive behaviour on-board. The most serious impediment is that most countries do not have laws that enable them to charge someone for an onboard incident of a foreign air carrier landing in their territory. Diplomatic immunity involving disruptive acts on-board has also caused victims to be angry and frustrated. Fair recourse of on-board disturbances is far from being realized.

In the case of the United States, the power to lay charges for crimes on-board aircrafts rests with the Federal Bureau of Investigation (FBI). Critics point out that it should be the airport enforcement units that have this power in order to ensure a more effective response when law enforcement is called upon to deal with these incidents.

Similar conflict existed in Canada prior to the privatization of major airports between the Royal Canadian Mounted Police and local police. This conflict has now been resolved with local airport police enforcement units responsible for responding to incidents reported by the airlines. As a result, the interface with the airline personnel is clearly defined, response times improved, and delays in investigation and follow-up reduced. Issues of varying jurisdiction in different countries can be very confusing for the operating crew when faced with an on-board incident since the decision to report such an incident is also affected by the understanding of what can be expected in terms of support upon arrival. This is especially critical when crews are in a foreign country and the time required for follow-up may well affect crew legalities, the airlines schedule, cost incurred and even a potential dismissal of the seriousness of the incident by local enforcement.

One such example is the United Airline captain who laid assault charges against an unruly passenger on a flight from San Francisco to Osaka on November 27, 1999. The F.B.I. forwarded a letter containing the charges to the Japanese authorities, however according to the captain, the Japanese police have not taken any action against the passenger as of early June 2000.

The responsibility lies with the airlines to provide their crew with this information and develop contingency plans for those incidents where support from local airport enforcement is weak or non-existent.

The emphasis of work to date is on improvements in three areas, legal refinements, enforcement handling and crew training. ICAO's Legal Commission launched a Study Group in 1998 to address this problem on an international scale. The air carriers have made few and then only sporadic efforts in creating public awareness leaving it up to the associations and regulatory authorities to voice their positions instead. The aspect of critical stress incident management for passengers and crews following a serious incident of passenger misconduct has been absent from the ongoing debate. This chapter examines what is being done in these areas and offers some suggestions and guidelines to achieve a more integrated and cohesive approach.

The models selected are intended to be helpful in clarifying:

- The connection between organizational issues and passenger misconduct;
- The interaction of four key elements affecting the outcome of the airline worker engaged in dealing with a passenger misconduct situation.

Critical Path to Disruptive Acts

Introduction

The larger regulatory, social, political and economic environment in which airlines operate affects organizations, their value systems and the priorities for decision-making. The examples given intend to identify some of these influences and their relations to latent and active organizational failures[1] to address the issue of disruptive passengers.

The regulatory environment is critical in achieving effective communication between the flight and cabin crew during an onboard emergency. The interface between these two groups is poorly regulated nor are there industry standards to that end (Chute, 2000). The necessity for clear communication processes and a commonly shared terminology to counteract mutual misconceptions and stress related deficiencies have been amply documented following a number of incidents. The disturbing fact remains with regulators hesitant to act on recommendations by the safety boards.

The issue of safety culture and its effects on accidents has been identified and raised by experts over recent years. The National Transportation Safety Board (NTSB) gave special recognition to this topic by organizing a Symposium on Corporate Culture and Transportation Safety in April 1997. Research produced by Chute and Wiener (1995a, 1995b, 1996) proposed an Information Transfer Model to understand the factors impeding the effective flow of information.

The principles of a well functioning safety culture[2] also apply to the effective management of the disruptive passenger issue. The airlines' organizational culture is crucial in defining the many sub-cultures that co-exist and make up the total picture. A corporate safety culture is one such aspect. As has become apparent in our research, those responsible for corporate safety were slow to face their responsibility in dealing with the less tangible nature of the disruptive passenger issue. One of the difficulties relates to the lack of a meaningful reporting mechanism and follow-up processes aimed at categorizing incidents for management analysis and action. Given the less tangible nature and also the more episodic nature of passenger misconduct at first, this is understandable.

To date there is no significant progress in harmonizing data gathering methodology and data analysis within the airline industry or by regulatory authorities. One of the approaches towards a common database has been initiated in 1999 through the efforts of the Transport Canada Working Group on Prohibition Against Interference with Crew Members. Due to the voluntary nature of this program established by Transport Canada, the national database suffers from a lack of submissions by the airlines, thus defeating its purpose.

Organizational Issues

Although we cannot hold airline senior management responsible for societal unruliness, airline executives have more control than other groups in taking the many steps needed to curb unruly passengers. They can influence lawmakers, revise training and procedures, develop safety-service synergy, and make corporate changes.

The focus here is also reflected in the modified version of Reason's Error Model in chapter 5 and the Interactive Model presented in a subsequent section of this chapter.

As a basis for discussion, three factors are of main interest to us in defining the critical path of organizational aspects enabling and possibly leading to disruptive acts by passengers. These factors are:

- Predisposing traits;
- Pre-conditions;
- Mitigating factors.

Figure 6.1 illustrates how these factors influence the occurrence of disruptive passenger acts.

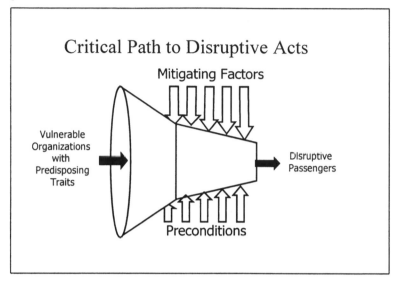

Figure 6.1   Organizational Path to Disruptive Acts

Predisposing Traits - The Short View

We are introducing a *'short view'* of predisposing traits first, followed by a *'long view'* in an effort to offer a brief historical perspective on some of the important developments over time.

Empirical data suggests that airline organizations vulnerable to disruptive passenger acts share a host of common traits. These are:

- Purely profit driven organizations;
- Conflicting corporate values and policies;
- Lack of a zero tolerance policy for passenger misconduct;
- Inexperienced organizations;
- Lack of 'violence in the workplace' programs;
- Lack of employee assistance programs;
- Lack of Critical Incident Stress Management programs;
- Lack of risk assessment;
- No formal reporting process for such incidents;
- No validated methodology for gathering and analyzing data;
- No clear identification of roles and responsibilities;
- No process for a joint effort with regulatory and legal authorities, law enforcement and international agencies;

- Corporate attitudes towards their customer service employees;
- Cultural attitudes impacting on the issue of disruptive passengers;
- Vague policies and procedures;
- Lack of a policy to keep passengers informed and frequently up-dated;
- Training based on regulatory minimum only;
- No or poorly designed preventive measures;
- No or poorly designed performance measurements;
- Product design without consideration for its impact on safety;
- Liquor policies;
- A lack of administrative deterrents for disruptive passengers.

Preconditions

Preconditions relate specifically to the individual risk factors of the perpetrator (usually unknown), and the victim (known), workplace risk factors immediately preceding an incident or inherent in the physical environment of the aircraft, and conditions related to the flight.

Mitigating Factors

Mitigating factors influencing the positive outcome of potentially disruptive passenger situations relate to a host of different aspects. Some are linked to the larger initiatives of the industry and regulatory authorities as well as legislators. Within the airline organization factors such as a zero tolerance policy, training and supportive fellow co-workers are positive forces in mitigating a potentially dangerous incident. The outcome is further influenced by the personal and professional make up of the airline worker, gender, and cultural compatibility with the perpetrator.

The International Air Transport Association (IATA) published its first edition of *'Guidelines for Handling Disruptive/Unruly Passengers'* in December 1998. Policies by British Airways, Cathay Pacific Airways, Canadian Airlines International, KLM Royal Dutch Airlines, Monarch Airlines, QUANTAS Airways Limited, Swissair, US Airways were re-printed together with additional reference material, with the FAA (see *Appendix E* for details) and REACT Ltd. being the main sources. Critics contend that the guidelines are predominantly aimed at responses to incidents in progress and short on preventive measures. No mention is made of airline marketing and its latent influence on disruptive passenger incidents through shortsighted service policies and product design.

The IATA Position Paper aims at seeking a broader implementation of the Tokyo Convention (1963) by certain Governments. As an industry lobbying association, IATA drives at obtaining amendments of domestic legislation in certain countries to tighten the legal safety net. This is an important precondition to prosecute disruptive passengers, including those involved in incidents on board foreign-registered aircraft.

Another step in addressing the global nature of the issue of disruptive passengers is the IATA Memorandum of Understanding (MOU). It targets countries where the scope of domestic laws implementing the Tokyo Convention is satisfactory. The MOU is intended to establish a blueprint of principles and procedures for handling disruptive passengers. This document endeavours to strengthen cooperation between airlines, airport authorities and police. The acceptance of this MOU will have to be achieved country by country respecting local legal, cultural, and political requirements.

These initiatives are very important on behalf of industry, however they take time and patience to lead to tangible results. Industry forums such as the IATA seminar on the topic of Disruptive Passengers in March 2000 continue to be instrumental in helping member airlines to share their experiences and identify best practices to date.

In the United States, the publication by the Air Transport Association of America (ATA), 'Airline Passenger Misconduct – A Guide to Prosecuting Federal Crimes Committed Against Airline Personnel and Passengers' is designed to help prosecutors to take legal action against passengers involved in an incident. The ATA claims that this document aspires to ensuring a more consistent approach for prosecuting offending passengers in the United States.

I have been unable to receive a critical review of this guide in time for the publication of this book. It is significant that it was produced without any input from the FAA and their legal department, a step that would have resulted in legitimising the concerted efforts of the ATA and the FAA.

Predisposing Traits – the Long View

Industries and organizations evolve in the context of their time with strong beliefs and values based on their successes. The aviation industry is no exception. Some of the major traits that have made an important and at times lasting effect on the fundamental relationship with the air traveler are rooted in the history of civil aviation. Its strong focus on technology has never fully addressed the air traveler as an important aspect of the safe operation of the aircraft.

The effects of introducing more sophisticated entertainment systems, and cabin design concepts (e.g. A3XX aircraft) such as the creation of onboard dance floors, shops and restaurants appear to further introduce potential safety hazards for the simple purpose of competitive considerations. The dominance of Marketing beliefs and values has overshadowed the relationship with the air traveler. It has not been a smooth ride.

Interactive Model

The following model is adapted from the work initiated by researchers from the Tavistock Institute of Human Relations and modified by Chappell and Di Martino (1998).

The model recognizes a variety of factors that could cause or contribute to violence at work or in our case, to interference with crewmembers. It has been amended to identify risk factors that are relevant to the aviation workplace however the main factors of the original model have been maintained. The model offers a framework for in depth and comprehensive analysis with the aim to uncover root causes and contributing factors in minimizing the phenomenon of passenger misconduct. The focus is on the individual and social risk factors presenting themselves in certain situations.

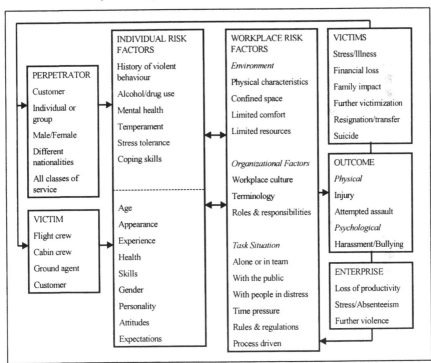

Figure 6.2 Interactive Model

We focus our attention on four key elements: individual risk factors associated with the perpetrator and the victim, workplace risk factors, and outcome of the abuse or violent act. Visualizing the unique and constant interaction between these elements is useful in better understanding the complexity of the issue; this in turn results in diagnosing the problem and make informed decisions to achieve improvements on a number of levels. The questions that need to be asked are:

- What elements under our control contribute the most to passenger dissatisfaction?
- What is the management level of commitment to reducing the factors triggering passenger misconduct?
- What airline organizational issues have a major impact on the management interventions of passenger misconduct?
- What are our existing and potential response capabilities of dealing with incidents to protect our employees and our company from stress, loss of productivity, absenteeism, and from further violence?

Perpetrator(s) – Individual Risk Factors

Constitutional laws are the major deterrents for developing a profile of passengers engaged in misconduct that would enable airlines to 'blacklisting' air traveler with a record of onboard misconduct. The Transport Canada Working Group on Prohibition Against Interference With Crew Members discussed this approach in detail. It considered: ... 'the rights of persons to carriage by air and privacy, the difficulties in determining objective criteria for denying carriage acceptable to air operators and the laws of different states, air operators' interests, the rights of crew to a safe workplace, practical aspects of regulating and administering such a mechanism on an individual air operator and national basis, the effect it may have on deterring unacceptable conduct, and initiatives taken by some air operators to deny carriage or otherwise take action against unruly passengers'.[3]

Industry security sources and enforcement agencies however confirm that males involved with alcohol abuse cause the most violent cases of on-board incidents. These findings are in in line with research results conducted in the area of violence at work (Chappell and Di Martino, 1998). It is noteworthy to point out that the International Labour Organization (ILO) has compiled a significant body of research on the topic of workplace violence worldwide and in many occupations. Recent incidents statistics for Ansett Australian Airlines (Loh, 2000) confirmed that males (87.1 percent) outnumber females (12.9 percent) in disruptive passenger situations. It underlines the aviation industry's need to address gender as a high risk factor associated with the occupation of crew and airline ground personnel dealing with the public. Looking at the limited database at our disposal the picture becomes more complex and reveals the need for further research.

- The perpetrator can be male or female, of any age, an individual or a group. Social status and education are also not distinguishing factors.

- Some of the more violent and abusive incidents reported by the media occurred in first or business class and involved citizens of privilege, diplomats and celebrities.

One such case occurred on EVA air in February 2000, when Hong Kong rock star Ronald Cheng, age 27, started to smoke and sing in First Class. When the cabin crew requested his cooperation to cease smoking and singing, he assaulted the cabin crew including some passengers, and threw himself on the floor shouting obscenities. He eventually was subdued and restrained.

Groups, especially tour groups, families, sports groups, and migrant workers, represent increased risks. Group behaviour follows a different set of dynamics; group norms may be very strong and make it difficult to fit in the structured cabin environment. Anonymity and group cohesiveness encourages emotional and rowdy behaviour (Deiner, 1980). One such event occurred on a Cathay Pacific flight from Johannesburg to Hong Kong in early February 2000. Reports indicate that a brawl broke out among a group of 30 South Africans during descent. The group, clearly divided into two camps, had been insulting each other for the duration of the 14-hour flight. The verbal harassment culminated in fistfights, injuring slightly four of these passengers. On arrival in Hong Kong, six people were arrested.[4] Group misconduct is a global trend, having resulted in some airlines taking special measures to curtail misconduct on their flights and protect their employees. Currently, airlines place little emphasis on investigating individual risk factors relative to the perpetrator although some factors are common to most air travelers.

Fear of Flying

Fear of flying[5] is a unique Human Factors' issue, intimately related to aviation. No other service industries have to deal with its consequences. It affects a large percentage of air travelers. For the majority of passengers, the aircraft environment is a source of stress. Various studies exist on the fear of flying,[6] also referred to in literature as aerophobia, aviophobia, and aeroneurosis, to name a few of the evolving terms. Fear of flying is frequently accompanied by additional fears, leading to a heightened state of fear of flying or in fact contributes to its manifestation. A study by Lufthansa (Krajc and Pausch, 1998) reveals that 55 percent of passengers suffering from fear of flying, are also afflicted with acrophobia,[7] 46 percent with claustrophobia,[8] and 4 percent with agoraphobia.[9] Fear of flying is considered a widespread ailment afflicting approximately 60 per cent of all airline travelers. A Boeing study (Dean and Whitaker, 1980) concluded that approximately 10 to 20 percent of passengers experience medium discomfort, while 5 to 10 percent experience severe symptoms by the mere thought of flying such as panic sensation and shaking.

Another study revealed that 41 percent of the respondents suffered severe anxiety while 51 percent suffered panic attacks (Byrne-Crangle, 1995).[10] It further confirmed that 65 percent of these fearful flyers used alcohol and drugs prior to and during a flight to combat their aerophobia.

The survey tool administered to the subjects, also categorizes flight related situations triggering aeroanxiety. Similar to the conflict continuum, the level of anxiety rises as an individual suffering from fear of flying. Typically, fear evoking situations are:

- Flight preparation, e.g. getting ready to leave, packing bags, media reports;
- Pre-flight, e.g. at the airport, getting boarding pass, clearing security;
- Crew directives, e.g. safety instructions;
- Take-off, e.g. engine start-up, aircraft acceleration, lift-off;
- In flight, e.g. unusual engine noise, vibration, turbulence;
- Pre-landing, e.g. safety instructions, deceleration, descent;
- Landing, e.g. touch-down, roll-out; waiting to dock.

Insight gained from this prepares cabin crews to be more empathetic towards air travelers and use their observation skills for early onboard intervention. Acknowledging the wide spread level of fear, understanding its origin and, reading the physical signs such as nervousness during boarding, sweating or freezing, shaking, not feeling well or dizziness are important symptoms that should not be left unattended. Cabin crew early intervention through personal attention and care can play an important role in helping these passengers to overcome or minimize their fears.

The boarding phase is crucial for cabin crews to identify and respond to requests for information, water, medication, pillows and blankets. Passengers engaging in this type of behaviour may be coping with their fears by seeking compassion and help. Demands made in an aggressive tone of voice and manner similarly are an expression of fear, although the passenger most likely attempts to mask his condition. In this situation, few cabin crews react appropriately by complying promptly and with grace. A third indication of someone experiencing fear of flying is a deliberate attempt to avoid responding to the environment, including to flight attendants. It is easy spot these individuals. They tend to shroud themselves in a blanket for the entire flight. This is a reaction more frequently noted in travelers from less industrialized countries, immigrants with little or no ability to communicate with the crew. Proactive cabin crews tend to these passengers despite conflicting task requirements during this highly pressured phase of pre-flight departure.

Anxieties evoked by cabin crew safety instructions pose some interesting questions: How do aerophobics respond to safety instruction given in an authoritarian manner compared to instructions given in a 'softer' communication style, e.g. using humour? Is there a difference in reaction to routine passenger briefings given in person rather than using safety videos?

The current emphasis on the public to recognize cabin crew authority possibly results in counter-productive interaction, simply because aerophobics' anxieties increase under pressure. In the end, it might be much more important how safety instructions are delivered than is generally assumed.

Fear of flying is not restricted to passengers; it also afflicts crew members. Some airlines,[11] having long recognized the benefit of offering treatment programs to their own workers, also make them available to consumers afflicted with aeroanxiety. The effectiveness of some of the major treatment programs have been the focus of studies by Byrne-Crangle 1995, Bornas and Tortella-Feliu (1995), Walder, McCracken, Herbert, James and Brewitt (1987). The academic interest in the fear of flying appears to coincide with the introduction of the jet age in the 1960s, and continuing up to the early 1990s. Since then, aviation research has been redirected into other areas.

Technology is finding its way into the treatment for fear of flying. The newest approach, virtual reality therapy, is competing with more conventional methods and championed by institutions such as the Center for Advanced Multimedia Psychotherapy in the U.S.A.[12]

Despite its usefulness, research into the phenomenon of fear of flying very rarely finds its way into flight- and or cabin crew training. This is a major shortcoming, since aeroanxiety is not something that disappears with the frequency of exposure to air travel, unless it is treated. Only a very small percentage of the public will undertake such remedial action. It is up to the airlines to gauge their services, and support cabin crews in detecting the symptoms, allowing for timely intervention.

Alcohol, Drugs and Medication

As discussed earlier, a common coping strategy to overcome fear of flying is alcohol and drug intake. There is however no research available that would uncover the possible connection of fear of flying, alcohol and drug abuse, and passenger misconduct. So far, incident reports give us few clues on what causes someone to drink excessively on-board.

Despite the cry of the overly zealous demanding a ban of alcoholic beverages onboard, the problem does not surface on every flight. In a seminar I conducted, representatives from 8 different airlines with different operating characteristics and from as many different countries confirmed that alcohol plays a part in causing disruptive passenger behavior, however it was not the major cause for all of these operators.

First, there is the issue of cultural differences and regional variances. Airlines operating out of countries were alcohol consumption is banned for religious reasons have little problems with their national customers. Other specialty airlines such as military transport operators have very few incidents related to alcohol. Airlines whose national culture accepts social drinking as part of its tradition are less likely to pinpoint alcohol as the major factor in disruptive passenger behavior.

Regional differences impact on the frequency of alcohol related disturbances as well. Air carriers serving communities where alcohol and drug abuse is part of the social condition experience a significantly higher risk than their large air carrier counterparts. For example, we analyzed incident reports of such a regional carrier in Canada for a period of 8 months. A total of 8 of the 19 incidents (or 42 percent) were alcohol related.

By comparison, the analysis of incident reports for one major carrier operating in the same country and for the same period of time shows a different trend. Out of a total of 19 incidents, 6 (or 31 percent) were alcohol related. Complicating matters is another look at the same period of 8 months the previous year. In this case, a total of 9 incidents were registered with only 2 related to alcohol.

One regional Canadian carrier servicing the far North, made the decision to discontinue the serving of alcohol on certain flights, due to the high number of passengers suffering from alcohol abuse. Flight and ground crews regularly discuss situations resulting in denied boarding due to passenger intoxication. Since there are no other carriers operating in that region, competitive issues are of no concern.

This very limited database offers some argument against advocates demanding a total alcohol ban on-board.[13] A more responsible management strategy would include considerations for regional and social predisposing traits, paid versus free bar, and preventive measures on the ground.

Enforcement statistics leave little doubt that the more severe forms of on board violence involve alcohol abuse. Clearly, airline management can examine their particular situation and determine what policies and procedures and, what product feature adds to a safe cabin environment for both its employees and customers.

Certainly the evolution of liability laws, especially in the USA, puts great pressure on the servers and sellers of alcohol. The quest of redefining accountability in responsible alcohol consumption on board an aircraft has just barely begun. No doubt major airlines' legal departments will analyze precedent setting lawsuits in the USA to assess their potential impact on future rule and policy-making. In one of the more spectacular cases, a major US air carrier was forced to settle a lawsuit with one of its customers who struck a motorcyclist after having consumed 10 to 12 drinks on a 2 1/2-hour flight.

Flight crews are now taking more proactive actions to prevent on board disturbances related to alcohol abuse. In April 2000, a joint decision by an American Trans Air captain and airline officials led to the cancellation of a flight scheduled to leave Indianapolis Airport for Las Vegas. The flight had experienced a delay of over seven hours. A significant number of the 165 passengers passed their time at the airport bar prompting the captain to be concerned with the potential of unruly and intoxicated passengers on board.[14] The question is, what measures did the airline take to look after their passengers during this lengthy delay that might have avoided the costly and drastic measures later on.

Drug abuse has become more of an issue as reflected in the section dealing with 'Successful Prosecutions', however information on the degree of its impact on abusive and violent passenger behaviour is difficult to establish in the absence of a substantial database. In our first analysis reviewing 91 cases reported to ASRS, only 2 indicated possible drug abuse versus 28 cases of confirmed alcohol abuse. The reasons for alcohol and drug abuse are complex. One reason, fear of flying, is particularly relevant to the airline industry.

Passengers on prescription drugs[15] may also risk certain ill effects during their travel. The attorney for the man, who broke into the cockpit of the Alaska Airlines flight, presented a 138-page report from his client's doctor. Reportedly, the man might have suffered a bad reaction to medication for his blood pressure causing him hallucinate. Bradley with no criminal record or history of mental illness, tested negative for alcohol or illegal drugs after the incident.[16] Subsequently, medical experts determined Bradley suffered from encephalitis, a viral infection causing inflammation of the brain.

Individuals requiring medication clearly have an obligation to speak with their physician regarding possible side effects and what alternatives to take before things go wrong when self-medication schedules may be interrupted because of travel.

Mental Health

Again, isolated cases being reported by the media resulted in severe injury to the flight attendant and perpetrator alike while the on board cabin atmosphere was saturated with general fear. Renee Sheffer is one of these victims. Other cases such as a Continental flight en route to Hawaii on May 17, 2000 resulted in a diversion to Guam. A male passenger became disruptive, kicking doors and throwing pillows with the cabin crew, assisted by some passengers eventually restrained him. He was taken for a mental evaluation prior to determining whether charges will be pressed against him.[17]

Another case involved a man returning on a Delta Airlines flight from Vietnam causing a major disturbance on board. According to reports, he had been on medication for schizophrenia for the past five years and had not taken his medication at the time of his flight.

An Australian woman traveling with her young son and nephew on a flight from Sydney to Francisco, complained about hearing voices, and began to act erratically. She started yelling and banging her head against the cabin panels. She resisted attempts by the cabin crew to restrain her, putting up a fight, biting and punching them before she was finally subdued. Upon arrival, she was taken to San Mateo County General Hospital for evaluation.[18]

It is clear that people with mental imbalances and on medication for their illness would be well advised to consult with their physician prior to planning any air travel to ensure that proper treatment continues or contingencies can be made to assist them to avoid a breakdown. The health sector and the aviation industry have an opportunity to work together on educating the traveling public, thereby assisting in a possible escalation of such incidents on-board.

Stress Tolerance

Passengers feel their individual stress tolerance[19] being put frequently to the test while using the aviation system. My research[20] indicates that 40 per cent of travelers engaging in misconduct experienced a number of problems including service failures prior to boarding a flight.[21] Problems include intoxication, unsatisfactory seat assignments including exit row seating, carry-on baggage disputes, long waiting periods between flights, delays, missed connections, lost baggage, missing meal, no meal choice, to name the more frequent ones. Others have special needs that the airline employee does not recognize or is unable to accommodate. The passenger manifest is designed to alert crews to such individuals, however this information is not always taken into account during the pre-flight briefing.

In a poll released in September 1999 by the Gallop Organization in the U.S., respondents were asked to rate 14 aspects related to passenger satisfaction with the airlines. The five leading factors causing air traveler dissatisfaction were in order of priority: legroom, in-flight food, ticket price, seat width, and airport parking. Except for airport parking, the four leading factors are under the direct control of the airlines. Airport parking was previously addresses in Chapter 5, Aviation System, where the airport design with its facilities provided grater details on the impact on traveler satisfaction.

When asked if they 'felt a sense of rage at the airlines or airline employees', approximately a third of the respondents confirmed that this was occasionally the case. Of particular interest was the finding that frequent flyers were less tolerant compared to occasional flyers. This is an aspect for airlines to consider in their analysis of policies and product design including the training of public contact employees.

Victim(s) Individual Risk Factors

Cabin crews run the highest risk of exposure to workplace violence in the line of duty. Although cases of abuse and assaults on flight crews have been documented, these are rare by comparison. Victims can also be fellow travelers, equally at risk from random violence, sexual harassment, verbal abuse and, physical assault.

The more disturbing cases involve sexual harassment of unaccompanied children and young female passengers.[22] There is no clear pattern emerging from the reports we have reviewed so far, although regional and cultural factors should be considered. Our preliminary findings concentrate on cabin crews as victims.

Age and Appearance

Empirical evidence suggests that young or effeminate looking flight attendants have greater difficulties in minimizing conflict situations, especially when they involve compliance issues or sexual harassment.[23]

Cultural and ethnic characteristics cannot be underestimated in their impact on dealing with disruptive passengers. Interviews and personal observations on board support the fact that gender, ethnic and cultural biases can have an impact on the diffusion or escalation of a conflict with a passenger. This is reflected in internal airline passenger surveys to the degree that findings indicate a bias against older cabin crews.

Uniform

The design of the cabin crew uniform is an important part in creating a certain image, to distinguish them from the public and, to underline their authoritative or service role. Marketing considerations affects the choice of design to a greater degree than concerns for practicality and personal protection in case of an aircraft emergency. In all the cases we reviewed where passenger non-compliance became an issue, flight and cabin crew authority was strongly resented. We have no data to compare the frequency of abuse and violence on carriers whose cabin crews wear less traditional airline uniforms.

Experience

In the case of cabin crews, our research indicates that younger, inexperienced workers are more at risk than their older and more experienced colleagues. These cabin crews are also more susceptible to suffer sexual harassment. These preliminary findings are in line with research conducted on a much larger scale and documented by Chappell and Di Martino (1998).[24]

Skills

Cabin crews with prior work experience in the hospitality industry are generally more successful in dealing with difficult passengers than their colleagues who come to the airline without such a background. Training in problem solving and conflict resolution should be a permanent focus, especially during induction training. Annual recurrent training plays an important part in reinforcing these skills to minimize the risk of becoming a victim in the workplace. Cabin crews from different airlines confirm their general perception that effectiveness of such training varies greatly. Conflict resolution skills are life skills, unless honed consistently and reinforced, little progress can be expected.

Gender

It appears that women have greater difficulty in obtaining compliance from passengers under duress. This is different in hijacking scenarios, where female cabin crews have proven to be more effective in dealing with the perpetrator(s). Cabin crew who by their appearance and demeanor create the impression that they are gay can cause homophobia when dealing with a disruptive male passenger as evidenced in some of the incidents I examined, resulting in heightened conflict rather than diffusing the situation. One classic response to passenger harassment related to sexual orientation: 'I am not a magician, I am only a fairy', has become legendary in cabin crew circles.

Personality and Temperament

My findings from a five-year study confirm that extreme leadership and interpersonal styles impact negatively on team performance and on customer satisfaction. Passengers identified predominantly two areas of concern, lack of responsiveness to their needs and failure to empathize when the situation required this approach. In all cases, the cabin crew involved was unable to use a flexible and engaging style to deal with the situation. Some examples given were: 1) ignoring passenger requests; 2) no assistance provided to parents with small children; 3) meal service very rushed, however pick-up of trays very slow; 4) no cabin crew insight after services completed; 5) cabin crew engaged in loud shop talk or on personal affairs, interfering with passenger desire to rest.

Attitudes and Expectations

Cabin crew and flight crew attitude and expectations play a key factor in their ability to successfully deal with a difficult passenger. Typical attitudes are expressed in the following quotes: 'I believe the public should be more informed about how to behave on an airplane and let them know what kind of authority flight attendants would have in case of disorderly and assault behaviour; I could not believe that a passenger would refuse to follow my instructions; suggest legislation giving airport police the authority to test suspect passengers for intoxication to determine fit for boarding; any and all people who allowed this incident to occur need education, re-training, and punishment.'

There is a strong underlying message that flight and cabin crew are a better class of people, and that they command respect. The gap between crew expectations and passenger performance is clear. These attitudes are dangerous and do little for blending server and customer performances to achieve a cooperative relationship.

Workplace Risk Factors

Aircraft Cabin Environment

The physical features of the aircraft as a workplace are very different compared to the traditional workplace referred to in literature. Simply put, the workplace is a thin-skinned metal tube that leaves the earth and rushes at great speed through the sky. It is utterly artificial and technically sophisticated. It requires crews and passengers alike to comply with mandated safety procedures from push back to parking at the gate.

Passengers frequently complain about the crammed aircraft seating, the uncaring personnel, the policies and procedures that show no regard for passenger sensitivities. The enclosed environment with an air pressure equal to 8,000 feet on jet aircrafts is very dry.[25] The air pressurization system removes nearly all humidity from the air, to which the body reacts. Restricting passenger mobility, scarcity of food and drink, too few washrooms resulting in lineups during long haul flights, are some of the factors contributing to stress and social tension in the cabin environment.

High-density aircraft seating configurations designed to maximize profitability add to heightened passenger tension. Being a well-adjusted passenger becomes a challenge, even to the most ingenious and experienced traveler when fellow travelers' personal habits and sanitary customs are offensive. Some examples of this are: slurping, belching, gum chewing, bobbing up and down while listening to music or tapping, putting feet up against chair backs, wiping one's nose on the back of hands or sleeves, body and foot odour, and manicuring one's nail; the list continues.

Aircraft Type and Length of Flight

The larger the aircraft, the longer the flight, the higher is the risk of incidents. Approximately 37 per cent of all passenger misconduct incidents reported to NASA's Aviation Safety Reporting System occurred on wide body aircraft, 19 percent on large and 34 percent on medium large aircraft respectively. These statistics are similar to my findings from major Canadian carriers, indicating the bulk of passenger misconduct occurs on long haul international flights and medium haul flights of over three hours.

Day Versus Night Flights

My examination of incidents on one major Canadian carrier revealed passenger outbreaks of abuse and violence occur more frequently on long haul and medium haul day (westbound) flights as opposed to eastbound and short haul flights departing at any time during the day or night.

The effects of the Circadian rhythm on alertness, a focus for extensive research on crew performance, also apply to the air traveler. Although medical evidence indicates that east-west travel is less disruptive to the circadian rhythms, a number of factors converge to create an unfavourable environment: Transit passengers have already crossed one or numerous time zones, disrupting their circadian rhythms and severely affecting their routine. They are tired, irritated and no longer able to perform adequately. For them the effects of environmental irritants magnifies.

Originating passengers have not had their circadian rhythms disrupted leaving them in a much more alert and normal state. The result is a demographic division between the already weary and the more energetic passengers. Both groups have specific needs for service but different in focus and timing.

Charter Flights

A cabin crew survey conducted at one North American charter airline indicates 23 percent believe they had suffered passenger assault, given the definition of the Canadian Criminal Code. Another 23 percent stated they had been involved in physically intervening with fighting passengers on board, while 34 percent indicated they had been verbally threatened and 24 percent had witnessed passengers tampering with on board safety equipment, including emergency exits.

Advertised and promoted as fun events, low fare charters set their customers up to false expectations. Conflict arises frequently in defense of one's space. People are ready to release their tension from work and home. Alcohol consumption is higher than on scheduled flights and inhibitions are lowered.

The scheduling of charter flights has a direct impact on intoxication prior to boarding. Hotel checkout times, requirements for extended airport check-in times and flight departure leave the passenger stranded at the airport for hours, with nothing to do. Airport bars are often the only place to escape to.

Aircraft maintenance problems often result in lengthy delays, especially if the carrier is unable to make aircraft substitutions.

Operational Irregularities

Delays due to mechanicals, weather, air traffic control, shift changes, crew legality (duty time) or availability are major irritants for the air traveler. A common complaint is lack of communication, incomplete and condescending information leaving the customer to feel like a hostage of the airline.

Other problems usually arise out of these situations, making things worse for the passenger: No access to food and drink or access to too much drink and no food. Boredom, deprivation of personal space, heightened anxieties, disruption to personal and business time tables with no or little control to make alternate arrangements.

Compliance

The most common situation resulting in potential conflict with the passenger is one where the cabin crew needs to obtain compliance to safety regulations. Cabin crews with little experience or lacking assertiveness feel uncomfortable because they expect resistance. Two predominant communication styles are used to overcome resistance, overly autocratic and overly timid. Cabin crew sex, ethnic and cultural background also appear to influence the successful outcome of a compliance conflict. As previously indicated, how these instructions are being delivered might well be the most important factor.

Organizational Issues

Terminology, Communication, Role Conflict

The aviation system as a centralized and highly structured environment uses its own terminology to communicate, thus creating a special language alien to the outsider. The passenger feels excluded when dialogue and understanding between the public and the crew should be encouraged.

Based on independent research conducted in the U.S.A, Canada and Germany between 1992 and 1997, passengers do not feel airlines effectively communicate with them. Typical complaints are that airlines are ignoring increased awareness and concern for passenger safety and quality service. The respondents share the view that airlines and their employees fail to provide sufficient safety-related information to achieve a more cooperative relationship with their customers.[26]

Cabin crews are expected to enforce safety regulations, to recognize individual customer needs and take responsibility to ensure that they are satisfied. This role of both enforcer and service provider causes at times conflict and confusion. Some cabin crews express their opinion, backed by regulatory regulations that 'they are here for safety reasons'. This implies, 'I am not here for service'. The type of relationship established between cabin crews and their customers can significantly contribute to an atmosphere where service is used as an aspect of diffusing tension rather than an interference with their more authoritarian safety role.

In September of 1999, after suffering major criticism about poor service by congress and air traveler advocacy groups, United Airlines announced it would deploy mobile computers at its gates to improve its handling of flight problems and to speed rebooking.

US Airways took a different approach by stating that it would create a consumer advisory board. The strong wave of consumer critique was prompted earlier this year when Northwest Airlines mishandled passengers stranded on its aircrafts for hours on end during the Detroit snowstorm debacle in January. It is working on an 'event recovery plan' to ensure delayed passengers are not held onboard an aircraft for more than an hour. Passengers should never be made felt like prisoners, should be given reasonable alternatives.

Policies and Procedures

Company policies and procedures are linked to deterring or enabling abuse and violence in the passenger cabin. A zero tolerance policy, backed up by a corporate safety net that draws its support from a wide range of organizational entities. Support from marketing, security, airport personnel, human resources, flight operations, in-flight services, labor associations, health and safety, and training is essential. North American carriers who have instituted such policies have achieved a significant change in employee attitudes towards their management. Some North American carriers have taken action to deter extreme passenger misconduct using temporary travel ban, loss of flyers' points as a means of deterrent.[27] This visible and published response gives much needed support to the employees who are faced with abuse and violence in their workplace. Due process required to safeguard against unfair punishment.

Changes in regulatory policies and procedures can also cause increased passenger tension. This was evidenced by the high number of incidents relative to non-smoking infractions and resulting passenger misconduct following the introduction of this policy in North America and on international flights.

Service Product Design and Procedures

The aviation community has a way of marginalizing passengers. This becomes visible in the manner airlines process passengers at the airport and again in the manner in which most economy services are designed. Security Procedures: One example is the incident involving Diana Ross the singer at Heathrow airport in September 1999. As she went through the security gate and the detector went off. She was then hand searched or more aptly described as padded down by a female security guard who Diana says got up close, feeling her breasts her buttocks and frontal thigh area. Apparently Ross took exception and asked if it was necessary whereupon the guard did it again. After clearing the check Ross went up to the guard and ran her hands over her breast with the comment this is how it feels to have this done to you. She then proceeded to board the Concorde for her flight to New York, however, before takeoff she was taken off the plane by police, detained for five hours, but not charged.

Marketing strategies such as providing free drinks to customers for competitive reasons should be weighted against the number and severity of complaints and incidents and revised accordingly. Passenger access to their duty free alcohol purchases or their private liquor supply during flight heightens the problem of alcohol abuse. Cabin crews are responsible to ensure that passengers comply with regulations, however inevitably this causes friction, especially when advertising and careless service design are contributing to the problem.

When designing food and beverage service, consideration has to be given to the time of departure, length of flight, and passengers in transit, as part of planning criteria. As competitive and profitability issues are foremost in the minds of airline executives, the cost of workplace violence and its impact on both the employees and other passengers needs to be quantified.

Task Situations

Alone or in a Team

Single cabin crew operations create different problems when compared to flights requiring more than one cabin crew. Flight Attendant Labour Associations have expressed their concerns to regulatory authorities, including in the sessions of the Transport Canada Working Group On Prohibition Against Interference With Crew Members. At this time of these discussions, no example could be cited where the safety of the operation was jeopardized due to passenger interference on a flight operated by a single cabin crew member.

When cabin crews work in smaller or larger teams, numbers do not necessarily equate with an ability to maintain social peace on board. There are documented cases that indicate an escalation of a problem with a passenger or a group based on inherent crew coordination and communication problems. This aspect has been addressed in greater detail in Chapter 4.

Dealing with the Public

Dealing with the public has become increasingly demanding, due to the generally high expectations by the passenger for quality service and personal attention. Blame for system failures is placed on flight attendants searching to explain situations over which they have little or no control.

A common cause for passenger anxiety is the lack of timely and meaningful information. Airline workers who deal with the public face passenger frustration and anger arising out of a general climate of stress and insecurity augmented by service problems. The airline worker has the disadvantage of not knowing what individual risk factors such as psychiatric disorders, alcohol and substance abuse afflict a passenger. Furthermore, he or she is not trained to deal with such cases, as is a caregiver in the mental health profession.

Time Constraints and Task Pressure

Work in the aircraft is driven by time constraints both in the cockpit and in the passenger cabin. The boarding and pre-flight departure phase is a very critical time for completion of mandatory routine pre-flight checks and safety related duties. It is also a time, when the passengers are anxious for reassurance when they first meet their caregivers, the flight attendants. This is a stressful phase for cabin crews and passengers, with one group demanding instant and personal attention, the other group with barely enough time to complete a multitude of tasks. Cramping of service features should be weighted against regulatory operational preparations.

Similar time constraints exist during pre-flight landing when the cabin has to be secured and passengers checked for seat belt, seat back and hand luggage compliance. Any additional problem presented to a flight attendant during these critical flight phases requires a very high degree of skill for satisfactory resolution. Certain cabin crew tasks will have to override individual passenger needs unless they are life threatening. For some individuals, this is extremely hard to accept and conflict escalates. Meal or bar service interrupted by turbulence for prolonged periods may be such triggers. Research however clearly reveals that timely, complete and candid communication with passengers will result in understanding and acceptance. There has to be some empathy shown to those who are left out.

Outcome

The outcome of physical injury or attempted assault on the victim cannot be underestimated.[28] The personal accounts of cabin crew having suffered trauma following an incident reveal the full extent on their personal and professional lives. The victim together with family and friends experience a fundamental change after such an incident. The psychological impact of repeated harassment and bullying can also be crippling to the victim and their professional colleagues. The results may range from stress to illness to financial loss to further victimization, transfer to another department or eventual resignation from their job.

The outcome of such events reaches beyond that of the victim. The wife of a Newfoundland man jailed for drunk and abusive behaviour on an Air Canada flight that diverted to Winnipeg expressed her feelings in an interview. Her concerns were palpitable; the possibility of her husband losing his job; looming welfare for her and her children; loss of company health coverage; the shame of it all.[29] Airlines need a comprehensive plan to deal with violence at work in order to protect employees and their customers in the multifaceted aviation enterprise.

Summary

Airlines control or influence the majority of the following factors:

- Victims and their individual risk factors, especially when determining selection criteria addressing experience, personality, temperament, attitudes, health and skills.
- Training is a key to developing skills based on the specific risks associated with an airline's operational characteristics, type of fleet and route structure.
- Circumstances conducive to passenger misconduct such as operational irregularities, over booking policies, service product design and customer compensation policies and procedures can be used as positive strategies to reduce risks.
- Environmental factors of aircraft design, particularly the degree of passenger comfort available on long haul flights.
- Organizational issues including workplace culture and the integration of safety and service roles are very important in achieving a better balance in the relationship between the airline employee and the customer.
- Task situations involving the design of service features at critical times during all phases of flight can cause conflict between a flight attendant's safety and service role. This is an important area to examine on an individual basis with a new understanding of marketing airline service.
- The organization's responsibility towards the outcome of incidents of passenger misconduct needs to be addressed in a comprehensive strategy to deal with violence at work.

One can argue that the airlines have no control over the perpetrator since there is nothing or very little known about his/her history of violent behaviour, alcohol or drug abuse, aerophobia, stress tolerance, and coping skills. On the other hand, airlines have a clear idea what situations cause major customer frustration. Complaints and incident data are a rich source for such 'situational intelligence' in determining what measures could be used to affect a reduction of risk. It is smart management action. The cost of failing to do so is high, and not too difficult to assess. The airlines have this information, however it might not be readily accessible and coded for easy retrieval through their budget centers.

Enforcement and Security

Airport policing is an integral part of the safety net in dealing with unruly, threatening, intoxicated or violent passengers at an airport prior to flight departure or after flight arrival. It provides the airport community and the flight crew with much needed support in preventing and following up on incidents of air borne crime.

As a result of the Transport Canada Working Group Against Interference with Crew Members, education for airline staff on the law, enforcement issues, evidence and court preparation became one of the priorities to address. In 1999, the Peel Regional Police Airport Division and the Ottawa-Carleton Regional Police Airport Policing Section published 'Unruly Airline Passengers, The Police Response, An information guide to airline staff in Canada'.

The brochures were distributed through the Working Group to all the stakeholders concerned, including airlines directly, and unions. They have also been sent out to many other stakeholders not represented at the Working Group. In addition, Peel Regional Police officers handed-out brochures personally to crews and ground staff at Toronto's Lester B. Pearson International Airport. Several thousand brochures were printed, including by the Ottawa-Carleton Police. The success of this publication speaks for itself. It has been reproduced by several other agencies and publications (like the ALPA news bulletin which itself has a huge circulation). It has also been adopted almost word-for-word into the Canadian Airlines cabin safety manual.

Despite the wide spread distribution and availability of this educational material, there are still cabin crew who are not aware of it. This is not unusual, what is unusual though is the overwhelming positive feedback this guide has received from the user community, including from airport police personnel in other jurisdictions. Due to the success of this guidance material its entire text has been reprinted with the kind permission of the two police agencies in *Appendix B*. This type of guidance material is very important for airline workers to have, and serves as a good example for similar material to be produced in other jurisdictions.

Partnership with Enforcement and Security

In a unique initiative by Transport Canada, the Director General H.H. Whiteman approved an interchange with the Air Line Pilots' Association (ALPA) to promote a better understanding of the role of civil aviation security in both organizations. Recognizing that security is not only an issue for the police agencies, airport management, or corporate airline security, but also for cabin crews to handle, a more comprehensive and inclusive strategy can only improve system performance.

Captain M. Sheehy, a senior pilot with Canadian Airlines International

Ltd., was the first flight crew member to be certified at the Regional Security Inspectors' Course in October 1998, while D. Smith, Chief, Inspection, Compliance and Physical Security for Transport Canada took part in ALPA's Security Training in the same year.[30] People with vision affect change; in this case it was Jean Barrette, Director, Security Operations, Transport Canada who was instrumental in seeing the benefits of breaking down traditional barriers.[31]

Solid partnership between the police and the airline staff takes time and effort to develop. The Working Group recognized the need to have a good working knowledge of each other's roles and responsibilities as one of the cornerstones for preventing and responding to passenger misconduct.

Captain Matt Sheehy, in his capacity of a member of the Working Group and as Canadian Security Coordinator for the Air Line Pilots Association, International (ALPA, INT'L), conceived the idea for a familiarization flight program sponsored by Canadian Airlines International with the Peel Regional Police being the first to participate in this innovative program. Malcolm Bow, a Sergeant with the Peel Regional Police and also a member of the Working Group was involved in the planning of this program. He summarizes his observations as follows:

> I was surprised at how busy the flight crew become in preparing for the flight and in some ways the business of flying airplanes and police work have more in common than you might think. It's clear that, like police work, a high level of communication, priority coordination and discipline are vital to accomplish a safe and secure flight. There is also certain intensity, an unspoken reality that at any moment you may be tested to the extreme, knowing that your actions will be closely scrutinized. This is something most police officers can also easily relate to.[32]

In the U.S., ALPA supported a local F.BI. initiative at Wayne County Airport (DTW) and the assistant U.S. Attorney, aimed to address the existing problems with Federal versus local enforcement response to incidents of passenger misconduct. In this test program known as CASE (Civil Aviation Security Enhancement), supervisory DTW police were sworn in as deputy U.S. marshals with limited federal arrest authority. Experience demonstrated the effectiveness of this program with an extension to Southwest Florida International Airport. Although more than twenty additional airports are exploring the initiative, CASE was recently disenfranchised for reasons ALPA is challenging.[33]

Sample Data

As part of the greater focus on passenger misconduct, the Peel Regional Police compiled statistics for 1998 and 1999. The data is divided into two major categories, on-board and pre-board incidents. The first table shows the total number of police responses for the period of 1998 to 1999.

The pre-board incidents category includes any occasion where police intervention was required to prevent a passenger from boarding an aircraft based on a range of offenses. This type of police vigilance and response is an important example for developing airport strategies aimed at keeping troublesome passengers on the ground while curtailing on-board incidents.

The onboard incidents category includes events where police were summoned to meet an incoming aircraft or to respond to an incident on an aircraft prior to push back.

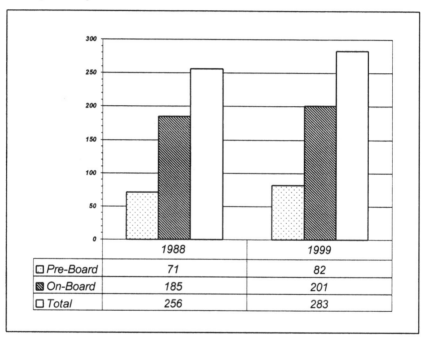

	1988	1999
☐ Pre-Board	71	82
▧ On-Board	185	201
☐ Total	256	283

Figure 6.3   Combined Incident Response (PRP)

The total number of times the police was summoned to intervene in pre-board incidents rose from 71 in 1998 to 82 in 1999. This represents an increase of 13.8 percent from the previous year. The combined total for police intervention during pre-board and on-board incidents rose from 256 in 1998 to 283 in 1999. This represents an increase of 27 or 10.5 percent.

Disposition of Incidents

The second set of figures shows the total number and disposition of police intervention for both pre-board and on-board incidents for the period of 1998 to 1999.

The disposition of incidents is captured by four sub-categories, namely arrest (Arrest), charges (Charges), police caution (P-Caution), and assistance only (Assistance). Some of the categories may be multiple, e.g. arrest, then cautioned, charged or released unconditionally. 'Assistance only' may mean assisting airline staff escort a passenger from the area or removing individuals under the Trespass Act or other legislation.

Figure 6.4 shows the disposition of pre-board incidents for 1998 to 1999.

Over this period, increases are noted in the number of arrests from 35 to 40, charges from 24 to 40, assistance from 18 to 27, while police caution decreased from 27 to 7.

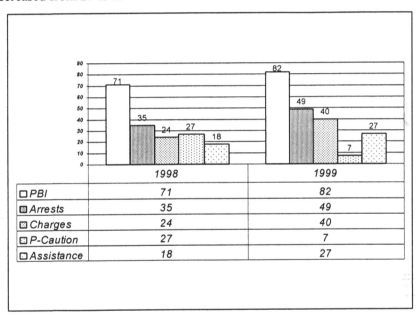

	1998	1999
□ PBI	71	82
▤ Arrests	35	49
▨ Charges	24	40
▥ P-Caution	27	7
□ Assistance	18	27

Figure 6.4   Pre-board Disposition (PRP)

Figure 6.5 shows the disposition of on-board incidents. A similar pattern is reflected as shown in the *figure 6.4*.

Increases are noted in the number of arrests from 53 to 54, charges from 41 to 69, assistance from 76 to 102, while police caution decreased from 70 to 37.

In summary, there is a noticeable shift in the way police responds to pre-board incidents in 1999 compared to 1998. There is a trend to decreased police caution and a marked increase in charges being laid and arrests.

Other enforcement agencies, such as at San Fransisco Airport reported a total of 111 on-board incidents over an 18 month period.[34] These figures are no cause for alarm when compared with the examples provided here covering a

12 month period.

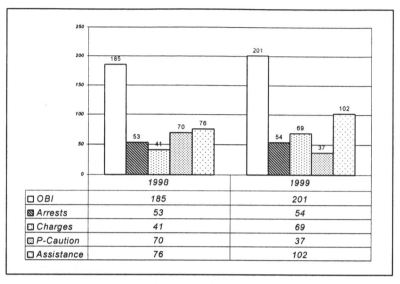

	1998	1999
□ OBI	185	201
▨ Arrests	53	54
▣ Charges	41	69
▨ P-Caution	70	37
□ Assistance	76	102

Figure 6.5   On-board Disposition (PRP)

Conclusions

These statistics have to be interpreted in the context of a passenger traffic increase of 4.1 percent (27,800,000 in 1999) over 1998 (26,700,000). The media, including advocacy groups simply focus on the increase of reported incidents year over year. In this case an 8.6 per cent increase of passenger misconduct is potentially an alarming situation for the traveling public.

In 1998, the likelihood of a passenger being involved in an on-board incident was 1 in 144,324. In 1999, it was 1 in 138,308 passengers. The traffic increase of approximately 1 million-passenger also increased the chance to be exposed to a passenger misconduct incident by 6016, consistent with the passenger increase of 4.1 percent. Since evidence is already pointing to long-haul flights and large airplanes incurring higher rates of passenger misconduct, a more meaningful context for analysis would be to relate the number of incidents to the number and types of flights operated in and out of an airport.

In summary, the situation for on-board incidents has remained relatively stable for the two-year period, a credit to the implementation of a number of initiatives over the last couple of years and their positive impact on social peace on-board and at the airport. A number of factors have contributed to a stable situation:

- Increased sensitization of Law Enforcement at airports across Canada to the potential impact of passenger misconduct on aircraft safety;

- Closer cooperation with Regulatory Authorities, associations and industry;

- Media coverage;

- Increasing public awareness.

Other noteworthy enforcement initiatives have been reported in the U.K. Police at Manchester Airport introduced a 'Disruptive Passenger Protocol' in cooperation with the airlines. The protocol confirms the police will prosecute every incident of passenger misconduct occurring on flights to and from Manchester. Airlines have agreed to submit a complete list of witnesses and ensure crew members involved are given the needed time off to complete their statements.[35] In July 1998, the Chief Inspector of Gatwick Police conducted a seminar for airline workers with the aim to make the aware of the application of criminal law to the aircraft and the role of enforcement.[36] The guidelines outlined for ensuring better cooperation between the airline workers and the police closely resemble those issued by Peel Regional Police.

What is not known however is the scope of incidents related to passenger misconduct towards their fellow passengers, particularly concerning unaccompanied children and young females traveling alone. This is an under-reported area with a special focus for cabin crew monitoring during flight.

Successful Prosecutions

Prosecution is an important aspect of deterring passenger misconduct involving assault, interference with crew and the safe operation of an aircraft. The reporting of such proceedings and their results in the media play a major role in public awareness and education. The summary provided is by no means complete, however serves as an example of the positive changes in the approach by the legal community and the media over recent years in following-up on high profile cases.

Australia
06-03-2000: Lawyer Diane Louise Hewens, 49 pleaded guilty in the Brisbane Magistrates Court to offensive and disorderly behaviour on Ansett flight 220 and fined $600.
Hewens admitted to having consumed alcohol prior to boarding the flight and being inebriated. She ordered more drinks on the flight and became disorderly, screaming abuse, hitting another passenger on the head and then kicking the seat in front of her. She was eventually subdued by the cabin manager with the help of a passenger and handcuffed.
After her sentence Hewett was unrepentant and blamed the airline for serving alcohol. Hewett reportedly has a history of previous convictions for assaulting police, stealing, serious alarm and affront, and offensive language.

Canada

12-09-2000: Michael Kenny, 51, was fined Can. $1,250 in Calgary provincial court after pleading guilty to three charges. He was caught smoking on an Air Canada flight from London to Calgary on December 5. Accused of being belligerent and grabbing and pushing a cabin crew member led to assault charges. His defence lawyer stated that the man had been drinking before and during flight and was taking medication for a back injury.[37]

01-07-2000: Prince George, B.C. A woman who took three smoke breaks on a one-hour flight from Victoria to Prince George was fined C$250 and sentenced to one day in jail. Jorgenson, 38, had pleaded guilty to tampering with or disabling an airplane smoke detector, a federal summary conviction.[38]

04-20-2000: Calgary, AB. John Malcolm Hughes, 46, was sentenced to 14 days in jail after pleading guilty in provincial court to a charge of property mischief and uttering threats on a flight from Toronto to Calgary on April 15. Hughes had consumed four beers on the flight before the cabin crew informed him that he would not be served further alcohol. This refusal prompted Hughes to shout obscenities for the next 20 minutes, causing other passengers to complain. Upon deplaning he uttered threats aimed at the flight and cabin crew.
Hughes lawyer argued that his client had forgotten to take his medication for controlling his anxiety and suffered from memory loss due to years of cocaine and heroin abuse.[39]

12-11-1999: Bedford, N.S. A Swiss executive Dolder, 39, was convicted of assault, sexual assault and for endangering the operation of a Continental Airline's flight on October 11, 1999 returning to Zurich, Switzerland from Newark, N.J.. Judge Bill MacDonald found Dolder not criminally responsible for his actions.
Dolder's lawyer introduced the testimony of two psychiatrists confirming his client suffers from a' bipolar mood disorder with psychotic features'. Dolder reportedly had packed his medication in his checked luggage. Thirty minutes into the flight he went into a rage, grabbing the buttocks of one female cabin crew member, hit a male co-worker in his face and continued kicking and punching, struggling for ninety minutes before being finally subdued.
The captain decided to divert the fight to Halifax where Dolder was taken into custody by the R.C.M.P. This was the first case in Canada where a person was charged with endangering the operation of an aircraft.[40]

11-18-1999: Robin DeGroot, 36, from the Netherlands, was fined C$4,600 plus court costs after pleading guilty to assaulting a cabin crew while drunk and unruly on a Canada 3000 flight from Amsterdam to Calgary on August 29, 1999. The incident resulted in diverting the flight to Churchill, Manitoba.[41]

08-07-1999: Steven Kille, 42, of England, was fined C$300 for assaulting a female cabin crew veteran of 35 years on an Air Canada flight from London to Edmonton. The incident took place when the cabin crew asked his wife to hold their infant child during a phase of turbulence according to safety regulations, rather than keeping the infant in the bassinette. Kille did object to the real or perceived brusque manner of the cabin crew member and slapped her on the back.[42]

08-05-1999: Whitehouse, 41, a Canadian official at an environmental agency in Geneva, was fined C$500 for public mischief and common assault. He became unruly after a few drinks and took a swing at a female passenger on a Sabena flight from Brussels to Dorval International Airport on August 1, 1999.

Singapore
08-14-1999: Richard Weeden, 34, from the United Kingdom, was sentenced to 12 months in prison and fined 360 Pounds for assaulting cabin crew while being intoxicated on a British Airways' flight from Perth via Singapore to London.

01-31-2001: Swarup Das, 27, a computer engineer, was sentenced to three strokes of the cane and one year in jail for sexually assaulting a woman on a Singapore Airline's flight from San Francisco in September, 2000.[43]

South Africa
06-14-1999: J. A. van Wyk was sentenced to five years in prison by a Kimberley Regional Court Judge. Van Wyk had reportedly refused to fastened his seat belt and paid no attention to the safety demonstrations and information provided on a South African Express Airways flight.[44]

United States
01-22-2000: Hung Cong Duong, 30, was sentenced to six months in prison plus five years of probation for causing an incident on a Delta Airlines flight on July 24, 1999. Furthermore, U.S. District Judge Howard McKibben prohibited Mr. Duong from flying on a commercial airline for the next two years and pay Delta Airlines US$4,200 in restitution.
Mr. Duong had been previously diagnosed as a schizophrenic and been on medication for the past five years. He was returning from Vietnam where according to reports his medication was confiscated resulting in his disorientation.[45]

12-14-1999: Celeste Keenen, 37, was sentenced to three months of house arrest, a US$2,500 fine and to perform 200 hours of community service for kicking a twelve-year old boy's seat causing him to hit his head while being trapped in his seat. U.S. Magistrate Joel Rosen ruled Keenen was guilty of assaulting a passenger on July 9 on board a Spirit Airline's flight from Ft. Lauderdale, Fla., to Atlantic City.
Her lawyer confirmed that since the incident Keenen has been diagnosed as manic depressive, and is now being treated for her condition.[46]

11-23-1999: Joe Luis Mendez, 35, was sentenced to two years in prison after pleading guilty to interfering with a flight crew resulting in diverting Southwest Airline's flight 923 from San Diego to Burbank Airport on January 12, 1999. Mendez has reportedly been ordered to stay on psychiatric medication after his release.[47]

10-23-1999: Frank Janicki, 37, from Belgium, was sentenced to four months in prison in U.S. District Court in Alexandria for verbally assaulting and intimidating a crew member on United Airlines' flight 951 on August 29 from Belgium to Dulles International Airport. Janicki was further ordered to pay a US$5,000 fine.[48]

United Kingdom
05-17-2000: Gerald Howard was sentenced to six months in prison after assaulting a seven-year-old girl on a Northwest flight in May 1999. Howard was accused of sexual contact with the girl.

10-28-1999: David Ansari, 33, was sentenced to two months in prison after admitting to swearing and shouting at cabin crew on a holiday flight from Palma de Majorca to Glasgow.[49]

10-15-1999: Nicholas Paul, 26, reportedly returning from a film shoot in South Africa on a flight from Cape Town to London, was sentenced to three months in prison by Isleworth Crocn Court for tampering with the smoke detectors and verbally assaulting passengers and cabin crew.[50]

08-21-1999: Alan Grave, 53, was sentenced to 15 months in prison after admitting to being drunk on a flight to Jamaica and indecently assaulting a female passenger. Reportedly the cabin crew knew about Graves' behaviour and had attempted to reseat him, however without success. Graves was already under the influence of alcohol prior to boarding the flight. He claimed shoulder pain and fear of flying prompted his alcohol abuse.[51]

07-22-1999: Neil Whitehouse, a 28-year old oil worker, was sentenced to 12 months in prison after refusing to turn off his cellular phone on a British Airways flight from Madrid to Manchester in September of 1998.[52]

07-07-1999: Kevin McGuggon, 32, was sentenced to two years in prison on July 6 after admitting to endangering the operation of a Continental Airlines' flight from the United States to Gatwick. McGuggon attacked crew members, butted fellow passengers while attempting to break into the cockpit.[53]

05-29-1999: Ian Bottomley, 36, was sentenced to three years in prison after found guilty to being drunk on an aircraft, endangering the operation of an aircraft and the safety of its passengers and affray on a British Airways flight 056 from Johannesburg to London on January 16, 1999. Bottomley had been drinking prior to boarding the flight and refused to comply with the cabin crew's efforts to curtail his twelve hour rampage, causing 30,000 Pounds of damage to the cabin interior. Bottomley denied all charges.[54]

03-11-1999: Robert Kimmet, 45, was sentenced to a 500 Pound fine after admitting to threatening a female cabin crew on a flight from Majorca to Glasgow in October 1998.[55]

Summary

A profile of the most severe cases leading to charges and successful prosecution shows the passengers are predominantly male with an average age of 37 years. (The two reports lacking identification of the age have been discounted.) Out of the 23 cases recorded, 18 involved males. Five males were identified as professionals in the higher than average income range. The three reported cases on female offenders averaged 41 years of age with no reference to their occupation except in the case of the Australian lawyer. The Hewens' case confirms the profile of a perpetrator in key aspects since she had a previous record of violence and according to the incident on the Ansett flight, also abused alcohol. Her gender deviates from the general profile and confirms that gender is no barrier to disruptive behaviour. Reports of the other cases do not make reference to the perpetrators' prior history of violence.

As a matter of interest, Statistics Canada recently released its report on domestic violence in Canada.[56] It was the first time men were included in this type of survey. The survey found that 8 percent of women and 7 percent of men reported they had been experiencing family violence. The report also indicated that men and women under 25 years of age are more likely to be victimized. How this type of pattern applies to the airline workers interacting with the public has yet to be researched.

Although sexual misconduct between consenting adults and sexual harassment of unaccompanied children has resulted in successful prosecution, young female passengers traveling alone are also at risk. The few incidents that have been reported occurred on long-haul flights. Sexual misconduct onboard does not fall under the general topic of air rage, however airlines must address this issue in order to protect the traveling public, and especially minors from harm while under their care.

Figure 6.6 provides a summary of the number of successful prosecutions by country.

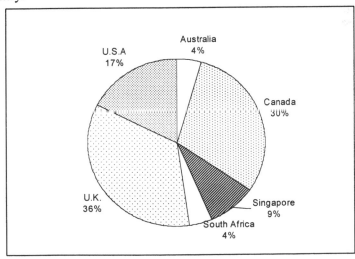

Figure 6.6 Successful Prosecutions (Overview)

Figure 6.7 depicts a summary of the factors that are not mutually exclusive and contributed to these incidents.

Intoxication contributed to 8 of these cases, medically confirmed mental instability (MHA) to 5 incidents, and opposition to compliance requirements 7, 3 of which were smoking related while 5 reports did not mention any contributing factors.

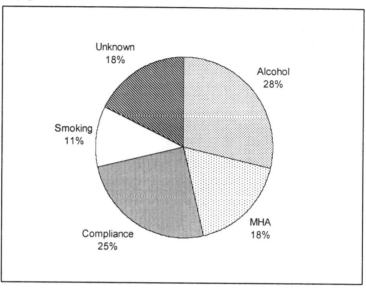

Figure 6.7   Contributing Factors

A total of 11 of these cases culminated in common assault, 3 in sexual assault, while 4 resulted in verbal abuse and threats.

These statistics are by no means comprehensive nor are they intended to be conclusive. They simply demonstrate to a very small degree the increased focus on successful prosecutions leading to greater public awareness and the more serious consideration given by the law in a number of countries around the world to incidents of onboard violence.

A Cautionary Perspective

Although the current trend in collaboration between the Canadian regulators, enforcement and the airlines is commendable, jurisdictions where the police lays charges rather than the prosecution, may result in the occasionally frivolous charge of a passenger, simply based on the statements of crew members who may flex their newly discovered powers under the protection of their airline and the law. The courts are overcrowded as is, and caution is in order to discourage charges as a result of limited fact finding prior to laying charges. The Crown is essentially a mouthpiece for the police compared to other jurisdiction such as in the United States where the police submits the information to the prosecutor. In the United States, a prosecutor will decide based on the evidence provided if the case should be pursued (Greenspan, 1987).

Righteousness may lead newly empowered individuals or groups to exploit their power until more experience has been gained, and the pendulum swings back to the middle ground. Previously, it was predominantly cabin crews who were at the receiving end of passenger malcontent, verbal abuse and even assault. There is now the potential for passengers who do not share the same commitment to civil onboard behaviour according to principles of righteousness, and do not ultimately interfere with the crew nor constitute a threat, to be the victim of an aggressive and hostile system.

A defense lawyer acting on behalf of his client may consider requesting access to airline policies and training programs in preparation for the case. A review would allow the defence to gather facts on the employee's record relative to past training, experience and skills in dealing with problems and conflict. Over-zealous, vengeful crew members could cause a passenger severe personal, professional and mental harm, not to mention major expenses.

As the legal arena concerning passenger misconduct evolves, the case of a wealthy widow launching a multi-million dollar suit against American Airlines for libel and punitive damages has been under way since 1996. The suit alleges that the airline filed documents with the U.S. Attorney's office that were selectively chosen, causing the plaintiff to suffer hatred, contempt and ridicule including being indicted on charges for interfering with crew members.

Public Opinion

The structured review of public opinion on issues relative to measures curtailing onboard passenger incidents is very much in its infancy. In a weekly survey conducted by AvWeb in the United States during the month of April, three questions were asked with the following results: 1) Can the airlines and/or government do more to defend flight crews from attacks by unruly passengers? Out of 670 respondents, an overwhelming majority of 94 percent felt that this compared with only 6 percent who answered no. 2) What's the single most important step that should be taken to protect flight crews from unruly passengers? Out of 673 respondents, 12 percent felt higher fines for passengers convicted should be levied, 25 percent favoured banning convicted passengers from all airlines, while 18 percent would like the flight crew to have access to defensive weapons (mace, stun-guns or similar devices). 3) Do you think the number of 'air rage' incidents are increasing, decreasing or about the same as always, just receiving more media attention? Out of 669 respondents, 65 percent felt that the number of incidents was on the increase compared with 34 percent that believed that the number was unchanged. No responses were noted for a decrease.

The results, although not statistically representative of the population at large, prompt the following observations: The questions were specifically related to the flight crew, however comments from the respondents extended to cabin crew issues. The results to the first question reveal a very strong feeling that neither regulatory authorities nor the airlines are taking sufficient action to assure the safety of crew members. The results to the second question point to a strong preponderance to punish the unruly passenger, not only by imposing higher fines but also to ban the convicted passenger from further air travel. The results to the third question reveal the influence of the media, indicating the majority believes that incidents are on the rise when current statistics do not support this fact in absolute terms.

This is in contrast to statistics maintained by the FAA. According to these records the number of cases has declined from 320 registered in 1997 to 266 in 2000, representing a 16 percent reduction overall.

In examining reader comments, a total of 43 support a total ban of alcohol service onboard while 26 feel it should be restricted. Greater emphasis of detecting intoxicated passengers on the ground and denying boarding is the third preferred strategy with 21 responses while others suggest breathalyzer tests to screen out intoxicated passengers. Airport bars are also highlighted in contributing to the problem.

Recommendations for crew training in defense and restraining techniques concentrate on defense techniques with only 2 responses highlighting the need for airline personnel to receive training in detecting troublesome passengers on the ground and conflict resolution skills.

A total of 17 respondents suggested the presence of air marshals armed plainclothes personnel onboard to deter and handle unruly passengers while another 2 suggested onboard video monitoring.

A total of 8 comments address the issue of service as a contributing factor to disruptive passenger incidents. The following expresses a passenger's point of view:

> Maybe if the attendants and airline people were a little more friendly and gave some small modicum of service to passengers, there would be less rage in the air and on the ground. I just flew with Continental and found the ground personnel arrogant and condescending with no clue as to what their job was. The seats were so small that I found myself hardly able to pass between the armrests. There is virtually no legroom (I'm 6' and about 230lbs., hardly a heavy weight.) On the redeye trip from Dulles to Houston, the attendants were surly at best and no one wanted to disturb them because of the rage that seemed to lie just below the surface. After cheerios and milk they just disappeared and sat back and talked among themselves. Maybe there is a valid reason to retire attendants by a certain age. The question that you should ask in a future survey should address these problems.
>
> There aren't any passenger problems since we don't fly every day. Service to passengers should be questioned more closely. I think that is where the main problem resides, with the airlines themselves. As for ticket sales and e-tickets,

forget about it and just have a drink and relax. That in itself is a black hole.

These comments from a veteran cabin crew member are sympathetic to the passengers' plight and take some issue with colleagues and airline management alike:

Sometimes partial responsibility for in-flight problems must be borne by the Flight Attendants involved and by airline management. In the 34 years I have been a Flight Attendant for American Airlines, I have seen situations escalate rapidly and unnecessarily because some Flight Attendants say or do things which add fuel to the fire. Despite many suggestions over the years, management has failed to provide conflict resolution training to the 23,000 Flight Attendants at American Airlines. I do not know if this is the case at other airlines.

As in all walks of life, there are some Flight Attendants who have a chip on their shoulder, they are unhappy in their own private lives, maybe they forgot that the passengers pay our salaries, possibly they are insensitive or uncaring or maybe they simply misunderstand the initial encounter with the passenger and become instantly confrontational creating a downward spiral of the situation. This would not give a passenger license to become verbally or physically abusive, however. I submit this scenario only because I have not see any mention of this aspect of the equation in reported incidents even though I have personally observed this very situation where a Flight Attendant, with the proper conflict resolution training, might have been able to influence a very different outcome. Because these incidents create problems affecting the safety of all the passengers and crew onboard a flight, the addition of conflict resolution training to our current safety training seems very appropriate and necessary. This suggestion is not the total answer but in view of the volatility of the in flight situations, it is an idea which should be given consideration. Possibly airline management has not done so because they fear corporate responsibility translating into lawsuits by Flight Attendants as well as passengers? I would appreciate hearing from you what other respondents have submitted since this is my livelihood.

These comments from a pilot reflect his views on cabin crew attitudes and the passengers' need to be treated with courtesy:

Many times the rage could have been averted by the crew, by treating the passenger nicely from the beginning with common courtesy. Too many times, we see a flight attendant with the attitude, 'We're here to save your butt, not kiss it.' I don't think this is the right attitude and provokes a response. Even as a pilot for a major airline, I feel it is my job, too, to kiss the butts of my passengers when I'm not busy saving it... after all, they put food on my table. Common courtesy from a flight crewmember can go a long way. However, in the instance that an unprovoked passenger exhibits rage on a crewmember, I am all for immediately landing, having the passenger removed from the airplane, arrested and prosecuted to the fullest extent of the law.

Based on the type of survey questions, the predominant emphasis of responses deal with responses to disruptive acts rather than prevention. Since the author did not have access to the demographics of the respondents, it is not clear how these results reflect the opinions of the population in general.

Under the Microscope

The following statistics offer a glimpse at data collected by four very different organizations. Methodologies and objectives differ, however to look at them in a complimentary way, leads to some interesting observations. The order of these examples directs reader's observations from the outside in. We will present data from a government research agency, enforcement, a training organization and a large airline. Also, time sensitivity dictates that their relevance lies in the past, like all statistics presented in this book. Useful, despite these shortcomings, they permit us to register shifts and movements in the various efforts to combat air rage.

In the first example drawn from the NASA Aviation Safety Reporting System (ASRS), the reader can see the dramatic increase in reports coinciding with airlines acknowledging that there is a problem and were prepared to examine the situation. These statistics cover incidents reported on U.S. carriers for the period of January to December 1998. The ASRS reporting system is voluntary with immunity provisions to protect any reporter's confidentiality. NASA has managed the ASRS operation for the last 24 years and established its reputation as a non-partisan research organization, unparalleled in its integrity anywhere in the world. Flight crews and air traffic controllers have used the ASRS database since its creation in 1976.

In the second example, additional statistic from the Peel Regional Police offer a closer look at the reasons for intervention, thus complimenting the statistics presented earlier in this chapter.

In the third example, we present a sample of an unidentified European air carrier. The airlines compiling internal statistics view their information as proprietary. By the nature of their enterprise they are not inclined to share their data unless there are regulatory requirements to do so, and then only in a manner that provides an overall reading protecting the identity of the airline. Concerns regarding liability, confidentiality and the potential impact on competitive aspects are underlying this lack of transparency.

In the fourth example, a report by a U.K. based training organization, SecuriCare, reveals the impact of class of service on the number of passenger misconduct. This approach is beneficial in identifying risks associated with other factors contributing to passenger unrest such as seating density and or personal attention given to passengers showing symptoms of discontent.

Within the airlines, labour and management follow their own processes of gathering data on incidents resulting in conflicting readings relative to the total number and categories based on the perceived severity of an incident. The same trend can be observed in the debate within the aviation system where labour and management use their statistics for their own purposes. Ultimately, there are no absolutes.

In the example of the Peel Regional Police, the enforcement unit at Canada's largest airport offers some hard evidence of how they have taken on the challenge to cooperate more closely with the airline community in Canada since the issue was raised through the Transport Canada Working Group On Prohibition Against Interference with Crew Members in 1998. In accordance with constitutional laws, these statistics do not disclose information that could lead to profiling the individuals that were charged.

Statistics concerning a profile of the victim are absent, except for the analysis conducted by the author in chapter 6, nor are circumstantial factors systematically addressed. SKYHELP is one organization with plans to do so. Research of organizational and cultural factors affecting the management of disruptive passenger incidents is only now beginning.

ASRS Statistics

The increased awareness by cabin crews of the availability of this reporting system has contributed to a major leap in reporting from this sector of aviation professionals. This is not unique since the reporting curve is also affected by the active promotion of a system, follow-up, reporter commitment and complacency.

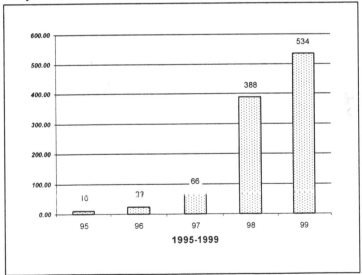

Figure 6.8   ASRS – Incident Reporting

Figure 6.8 shows the dramatic increase over the period of 1995 to 1999. NASA has done so with great success and with the support of various aviation organizations, including labour associations.

These efforts are paying off. Overall, the total number of incidents submitted to the ASRS has increased during the same period. Incident reports by cabin crews still constitute a small percentage when compared to the total incidents reported. Compared with internal airline data, of course these numbers are negligible, especially with the volume of air traffic in the U.S. The total number of incidents by one international air carrier for a one-year period was 371.

Figure 6.9 identifies contributing factors to incidents of passenger misconduct for the period of January to December 1998. Reports on passenger misconduct filed be cabin crews to ASRS fall into five major categories: alcohol or drug-related events; possession or use of potentially hazardous materials, devices, or substances; and bomb threats or hijacking attempts, and miscellaneous events such as uncooperative or unstable behaviour.

The ASRS analysis does not provide for a more detailed categorization of minor or more serious events such as interference with cabin crew duties or violation of federal aviation regulations. These statistics are however interesting when compared with the Peel Regional Police statistics, identifying the reasons for their intervention. It should be noted that the categories in the ASRS data are not mutually exclusive. As evident from the ASRS data, alcohol is the leading factor contributing to passenger misconduct, although there is no further breakdown to identify the number of incidents involving assault.

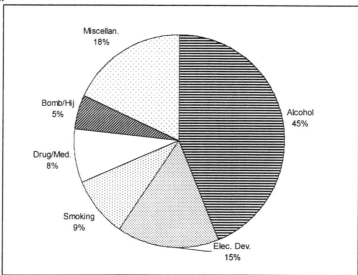

Figure 6.9   ASRS - Contributing Factors (Jan-Dec 1998)

The second highest contributing factor to passenger misconduct is the use of prohibited electronic devices. Clearly this is a reflection of the North American culture and its increasing dependency on electronic communication and entertainment devices such as cell phone, laptops, wireless palm devices or headphones.

Already regulators are facing growing concerns regarding the increased number of automobile accidents caused by the drivers' use of cell phones. The enormous marketing success of these devices of convenience however is causing a new set of problems. Their indiscriminate use is affecting road safety, while health concerns are also on the rise. The public demands the cellular telephone industry to disclose information on the amount of radiation that enters the user's head when these wireless phones are used.[57]

The proliferation of electronic devices is of concern to the aviation safety experts, and the public is confused about the rules on this issue applicable to aviation. The reported number of cell phones in the U.S. is approaching 90 million, growing at 30,000 new customers every day.[58] Although there is no hard evidence that these devices could cause interference with aircraft instrument landing or global positioning systems, experts presented their views at a House Transportation subcommittee hearing to Congress in July 2000. While a resolution of this issue cannot be expected soon, flight crew associations and industry support current standards.[59]

The third highest contributing factor is smoking on board although air regulations forbidding banning smoking have been in effect since the mid-1990s.

The fourth highest contributing factor is drug or medication induced behaviour.

Peel Regional Police Intervention – On-board Incidents1998-1999

The Peel Regional Police currently uses a total of 5 categories compared to 6 used by NASA's Aviation Safety Reporting System (ASRS). While the ASRS is voluntary, the statistics compiled by the Peel Regional Police are 'involuntary', and therefore provide a factual measurement of police activities related to passenger misconduct.

The on-board category includes incidents where police was summoned to meet an incoming flight or to respond to an incident on an aircraft after passenger boarding had started in preparation for flight departure.

The numbers do not include incidents involving passenger misconduct occurring after flight arrival, e.g. in the baggage area, Customs and/or Immigration halls.

A total of five categories are used to determine factors associated with on-board incidents (OBI): Unruliness and intoxication, mental health (MHA), assault, smoking, and other.

Figure 6.10 shows comparative data over a two-year period for 1998 and 1999.

As indicated earlier in the section on *Sample Data,* the apparent increase in police intervention is insignificant and reflects the 4.1 percent increase in traffic overall at Lester B. Pearson's International Airport.

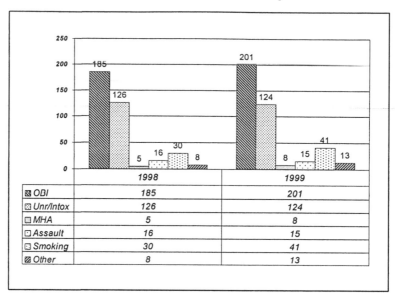

	1998	1999
▨ OBI	185	201
▨ Unr/Intox	126	124
▢ MHA	5	8
▢ Assault	16	15
▨ Smoking	30	41
▨ Other	8	13

Figure 6.10   Intervention On-board (PRP 1998-1999)

Figures 6.11 and 6.12 shows the distribution of contributing factors to on-board incidents in 1998 and 1999 respectively.

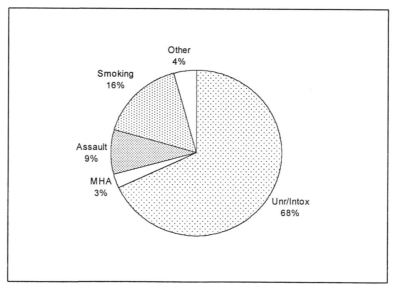

Figure 6.11 Contributing Factors On-board (PRP 1998)

Figure 6.12 shows the distribution of contributing factors to on-board incidents in 1999.

Incidents related to assault and intoxication show a small decrease while smoking related incidents show a small increase.

Factors such as airlines adding flight frequencies on international routes may have some influence on changes in such patterns although there is insufficient data to support such a hypothesis.

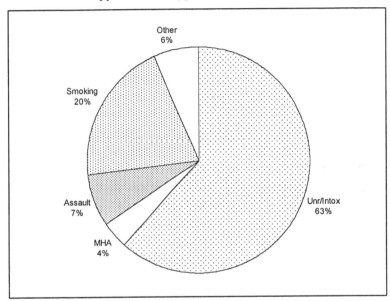

Figure 6.12 Contributing Factors On-board (PRP 1999)

Although the categories used in the analysis of the police data and those of the ASRS are not compatible, some areas point to some interesting discoveries.

Alcohol: As in the ASRS statistics, alcohol is the leading factor. Comments from other enforcement agencies at other airports also cite alcohol as the most frequent factor in incidents leading to assault.

Smoking: This is the second leading factor compared with the ASRS results placing fourth.

Assault: Forms of assault rank as the third contributing factor. There is no equivalent category provided by the ASRS statistics.

Other: This category ranks forth compared to the miscellaneous category ranking second in the ASRS statistics.

Mental Health Act: As in the ASRS statistics, this category ranks fifth in terms of frequency.

The two categories of electronic devices and bomb threats are not used by the Peel Regional Police in contrast to the ASRS statistics.

Observations

There are fundamental differences in the methodology used to collect the data presented above. In the case of ASRS, the reporting is voluntary, while the data produced by the Peel Regional Police is involuntary; one collects system data, the other localized data. The total figures for each one of the two systems also show the effectiveness of one versus the other.

Figures from these two different organizations, one located in the U.S., the other in Canada, point to some apparent cultural factors driving the divergent methodologies. The stereotypical offender as per the ASRS data overindulges in alcohol, needs his electronic gadgets, his cigarettes and medication. The stereotypical offender as per the Peel Regional Police statistics also overindulges in alcohol needs his cigarettes, resorts to occasional assault with a small percentage falling under the Mental Health act.

What is particularly striking in the Canadian example is the zero reading of improper use of electronic devices. Curious as this may be, although the widespread use of cell phones while driving has become an issue for Canadian regulatory authorities as well, the same impact cannot be observed onboard. This leads to further questions: Are these incidents a function of differences in consumer behaviour linked to demographics, to cell phone dependency and communication needs, the characteristics of a more benign aviation system or a function of differences in cabin crew training and skills?

Air traveler dependency on electronic devices in the U.S. is obviously a significant factor. In light of consumer complaints issued monthly by the DOT[60], this is not surprising since delays are the leading cause of passenger dissatisfaction and can have a substantially negative impact on business plans. Regrettably, none of these statistics provide us with the needed situational intelligence to substantiate our conclusion.

Similarly, in the cases of the Peel Regional Police statistics, a breakdown by route and type of airline operation would be of interest to lead to a cross-reference to our findings, identifying international flights as contributing to a greater degree to onboard incidents compared to domestic flights.

An increase or decline of voluntary reporting should not be interpreted as a realistic reflection of trends relative to the frequency of occurrence. Given options for reporting, the reporter's perceived value of a reporting system will have some bearing on where the raw data is funneled. For example if an airline's internal reporting system is meeting employee needs, the likelihood of using this system is greater than the rout of anonymous or duplicate reporting.

Airline Statistics

The following example illustrates the type of categories one large European air carrier has developed in combating passenger misconduct. The data presented covers the period from March 1998 to April 1999. During this time, the airline recorded a total of 371 incidents, more than one incident per day affecting one of their flights. Passenger incidents reported fall into three major categories:

1. Type of incident (five sub-categories)
2. Contributing causes (seven sub-categories)
3. Action taken by crew (four sub-categories).

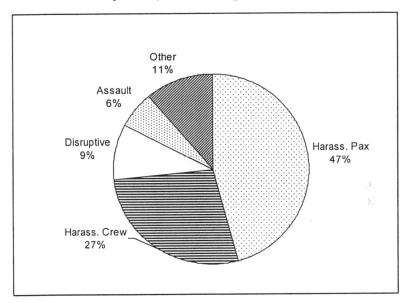

Figure 6.13 Types of On-board Incidents (Airline Sample)

This airline has included a separate category for 'harassment of other passengers'. It enables the organization to measure to some extent the degree of social risk onboard, and possibly address the impact on marketing strategies as well. Although the method for counting these incidents of passenger harassment has not been disclosed, the fact that they are being registered opens new avenues for investigation. Some of the questions to explore are: What is the impact on the victim, on passengers in the immediate vicinity witnessing this event? Are they secondary victims? If so, should counseling be available? Does this experience influence their future purchasing decisions with the airline? The social costs have not been explored to date. What if any, are the legal ramifications for the airlines?

Assault cases account for 6 percent. By contrast, this indicates that 94 percent of the incidents involved some form of verbal abuse and threats.

Figure 6.14 shows the distribution of contributing factors resulting in passenger misconduct. As a matter of interest, only 26 percent of all incidents recorded identify contributing factors.

The catergories are similar to the ones established by ASRS and the Peel

Regional Police however provide details of specific importance to the airline operation. Offences against non-smoking regulations are further defined by introducing two categories, one recording smoking in washrooms the other non-compliance in the passenger cabin. This allows for measuring the degree of safety risks specifically related to the potential of washroom fires.

It also might be beneficial to determine the distribution of smoking related incidents by route. It stands to reason that in countries where non-smoking legislation has been introduced for some time, a positive effect on reforming such behaviour would also be registered by the airlines.

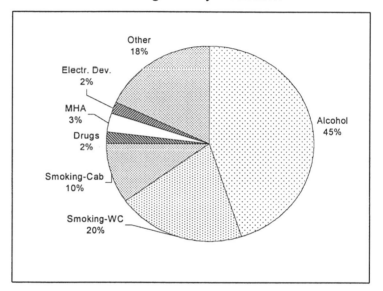

Figure 6.14 Causal Factors On-board Incidents

Looking at trends in the U.S., passenger misconduct involving smoking currently accounts for approximately 8 to 10 percent of all reported incidents. In Canada, smoking related incidents ranged from 20 in 1998 to 16 percent in 1999.

By contrast, one major Asian air carrier reports non-compliance with smoking regulations took a major leap after being introduced in March 1999, accounting for 21 percent of onboard incidents. No doubt, the period of transition for passengers adjusting to this new legislation will take several years, while the impact cabin crews face during this time is significant.

The combined percentage of incidents involving alcohol and non-smoking violations amounts to 75 percent of all incidents. This leads to the question, are these incidents a function of demographics, the psychological profile of the perpetrator, the onboard service environment or the aviation system? Obviously, there are no simple answers. Apart from the addictive nature of tobacco, alcohol consumption has been identified as one major strategy for passengers dealing with fear of flying. Since research on the fear of flying also indicates the degree of the affliction in air travelers, special consideration for airline service policies and focus on cabin crew training follow.

Figure 6.15 shows the outcome of actions taken by crews, using four major categories. Severe crew response accounts for approximately 47 percent of all reported incidents.

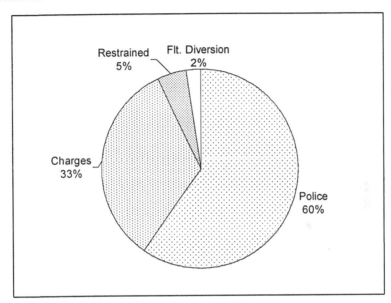

Figure 6.15 Crew Response

It is very important for the airlines to track what actions crews take in response to severe passenger misconduct events. From the victim's point of the view, follow-up on the results of legal actions taken against the perpetrator provide a critical step in dealing with any post incident trauma that might have been suffered. Airlines who regularly provide their workers with internal statistics of this nature make significant gains in worker moral. It demonstrated that the organization supports and appreciates the difficult actions their employees must occasionally take to maintain social peace and protect the safe operation of the aircraft.

This type of data gathering allows this airline to more readily measure the impact of passenger misconduct incidents in terms of direct and indirect costs. Out of a total of 371 incidents, 104 resulted in police interventions, 58 charges, 8 occasions of passenger restraint, and 4 incidents led to flight diversions. Costs based on the actions taken by the crew can be established. This is an important step in changing organizational awareness leading to reforming counter-productive attitudes towards their front line workers and institute critical incident stress programs to assist in dealing with the aftermath.

SecuriCare Statistics

Figure 6.14 shows statistics compiled by SecuriCare and offer some evidence to the benefit of including class of service as a worthwhile category into any system monitoring passenger misconduct incidents.

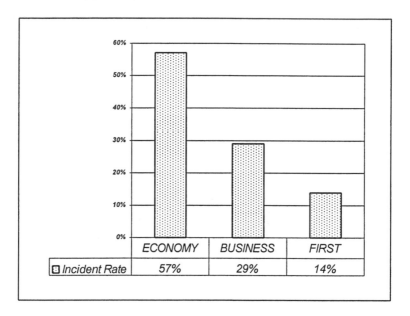

Figure 6.16 Incidents by Class of Service

Although SecuriCare avoids making any conclusions, critics of aircraft configurations and their impact on passengers' physical comfort are not surprised. Again, these findings lead to more questions, e.g.: What are the trends by aircraft type and configuration, on charter flights with high-density seating, compared to schedule air operators with improved seat configurations? In summary, this type of examination could help airlines to become aware of the choices they make for the sake of profit and lead to better decision-making.

Observations

The different examples used in this section show both similarities dissimilarities in how incidents are categorized. Some of the categories are useful to achieve an approximate comparison and compliment each other. On closer examination, all raise new questions that deserve to be explored. There is effectively no consistent method for monitoring passenger misconduct and analyzing such incidents. The impact of demographic and socioeconomic differences (as two of a broader range of factors) on the frequency of passenger misconduct is important to establish. New questions lead to new answers; they create focus.

Although some progress has been made over recent years to gather data, the benefits of a detailed categorization such as developed by Loh (2000) undoubtedly would lead to a much clearer understanding and improved management.

Victim Initiatives

SKYRAGE, an organization I introduced in chap.1 is a private initiative of Renee Sheffer and her husband after Renee suffered a severe onboard assault. The organization has expanded its focus and influence in creating awareness worldwide. Their website is predominantly an information resource to interested parties touching on all aspects of this complex issue. Since its inception, the format and content of this site have matured and reflect the increasing understanding of the multifaceted strategy needed to achieve results. Renee and her husband Mike have long passed the stage of anger and bereavement. They have been invited to speak at a number of aviation conferences and play an important part in maintaining focus on this issue. They exemplify how victimization can lead to a successful conclusion and affect change in a system that needs to be further sensitized.

Similarly, SKYHELP is a major private initiative by Susan Howland, a victim herself, who through her experience saw the need for better access to trauma counseling following a serious onboard incident.

Consumer Rights

The passenger rights movement in the U.S. is temporarily on hold, awaiting the assessment of voluntary measures carriers committed to at the end of 2000. Talks to make passenger rights a legal requirement continue while further pressure mounts through the introduction of the Air Travelers Fair Treatment Act July 19, 2000[61] and a proposed bill that would give the Department of Transportation wide powers to protect consumers in view of merger talks between United Airlines and US Airways.[62]

The European Commission has launched its passenger rights initiative in the summer of 2000. Largely modeled after the U.S. program features additional aspects such as standards for airport design, and cabin conditions on passenger health.[63] Section 62 of the Communication From The Commission To The European Parliament and The Council is of particular relevance to changing attitudes towards health issues ascribed to cabin conditions and measures dealing with 'air rage', and states:[64]

The responses to the consultative document showed wide agreement that the extent and nature of any problems had first to be determined, by examination of existing research and further work if needed. The Commission intends to set up expert groups to scrutinize existing research and draw conclusions on risks for health, in co-operation with interested parties. It will also explore whether Community measures could be taken to deal with disruptive behaviour by passengers (air rage). When considering possible Community measures in the above fields, the Commission will take full account of the principle of subsidiarity.

Countermeasures 1997-2001

In 1997, with the support of the Air Transport Association (ATA), Captain S. Luckey of ALPA was instrumental in organizing the first International Conference on Disruptive Passengers in Washington, D.C. This event surpassed his expectations of attracting just a small group of interested parties, when over 400 people showed up. Captain Luckey has continued his quest and to date addressed already 32 enforcement agencies in speaking engagements across the United States. In his own words: 'It goes a long way in achieving improved understanding and support.'

Acting as a watershed event, this conference officially launched more concerted efforts in combating air rage. As one example, ALPA co-sponsors its International Aviation Security Academy (IASA) with Embry-Riddle Aeronautical University, another is the Aviation and Security Management Certificate Program offered by the Aviation Institute, George Washington University.

A combined strategy of implementing airline zero tolerance policies and procedures with more effective legislation, such as in Canada, the United States, the UK, and Australia appears to have resulted in incidents of passenger misconduct leveling off. A multi-pronged strategy including awareness campaigns for the public were launched as previously mentioned in Canada on June 8, 1999. The first of its kind in the world, posters for display at airports and pamphlets in the form of ticket stuffers provided guidance to passengers.

In early April 2000, the FAA reauthorization bill responded to the demands of crew associations and the ATA by increasing maximum fines from US$1,100 to US$25,000 per incident against passengers convicted of threatening or assaulting crews or endanger the safety of the aircraft or passengers.

IATA reported on the progress of its initiatives at the Seminar On Disruptive Passengers in Geneva on March 23, 2000. The member airlines have adopted a model Memorandum of Understanding between carriers and airport authorities. Protocols similar to those of the Canadian airport enforcement agencies such as their counterparts, the Gatwick Airport's Disruptive Passenger Action Group, served as examples for other airports to follow.

Based on the IATA official press release of March 27, 2000, incidents of serious disruptive passenger were now limited to approximately 1 in every 4 million passengers. The press release does not provide any clarification regarding previous figures nor measurement methodologies nor time frames to establish a basis for comparison.

In October 2000, IATA in cooperation with ACI addressed the problem of unlawful interference in a symposium with aviation security experts in an effort to obtain input in developing international policies.

On June 20, 2000, the Working Group on Prohibition Against Interference with Crew Members presented its final report to the Commercial Air Services Operations Technical Committee containing eleven recommendations.

Transport Canada has accepted the recommendations that include proposed changes to the Criminal Code of Canada and the Labour Code, amendments to the Canadian Aviation Regulations, a tracking system of incidents, and continued efforts to maintain and expand an awareness campaign for the education of airline workers and the traveling public. As of March 2001, a tracking system is in place, however ineffective due to its voluntary nature.

On February 15, 2001, Transport Minister David Collenette introduced an interim measure, requiring 'that cockpit doors be locked during a flight when an incident or threat to safety arises due to unruly, abusive or dangerous passenger behaviour. This new measure will affect all major airlines in Canada.'

The reaction from ATAC and pilots was mixed. The pilot in command is legally responsible for the flight and needs a requisite amount of authority. Infringing on the captain's authority in this manner is perceived as diminishing the pilot's responsibility. Fuelled by the lack of consultation with industry and labour prior to the announcement, backroom discussions are now underway.

The media coverage too has become more thorough in reporting on key events with increased emphasis on providing follow-up on major incidents that have resulted in successful prosecution and are broadcast world-wide.

Air Travel and Health.

Some experts have warned about the health effects of low pressure, seat pitch and of cosmic radiation and about the risk of transmission of disease. Focusing on air travel and health, the House of Lords called on the Select Committee on Science and Technology to investigate a wide range of issues affecting passengers' and crew members' performance. Growing evidence from the medical community acknowledges the altered physical environment on-board compounds passengers' predisposing health risk factors.

The Fifth Report published November 15, 2000, provides a detailed review of issues and recommendations with far-reaching implications for the air transport system for a new direction in air traveler welfare.

Acknowledging the changed demographics of air travelers, especially in the sizes, ages and health conditions over the last 40 years, the CAA has launched a new research project. Its aim is to investigate the effects of these changes on the reduction in mobility after long-haul flights to ensure emergency evacuation requirements are adequate.

International Trends

Progress concerning the prevention of disruptive passenger incidents is closely linked to the advancements on several levels as mentioned earlier. Again national legal systems and culture are reflected in the approaches taken. As an example, the three major air carriers in Japan have issued a statement in April 2000 warning their passengers that they will be restrained should they threaten others onboard or refuse to comply with cabin crew requests.[65]

This warning came after releasing statistics that incidents of unruly passenger behaviour had more than doubled in 1999. The total number of 330 incidents included intoxication and sexual harassment. National culture with its resulting views on women in general and especially in a service role combined with previously identified risk factors have an impact on the type and frequency of these incidents.

The International Transport Workers' Federation (ITF) on behalf of its more than 200 civil aviation unions and representing more than 700,000 airline workers first presented its concerns relative to onboard passenger violence at the ICAO Assembly in Montreal in October 1998. In 2000, the ITF organized its first International Campaign Week for Action against Air Rage from July 3 to July 7, 2000. Activities at major airports around the world and included in the U.S.: Chicago, Fort Lauderdale, San Francisco, Boston; in Argentina, Buenos Aires; in Canada, Montreal; in France, Paris; in Germany, Frankfurt; in Japan, Tokyo; in Mexico, Cancun, and Mexico City; in Nigeria, Lagos; in Norway, Oslo; in Sweden, Stockholm; in Switzerland, Zurich; in Taiwan, Taipei; and in the U.K., London.

Cabin crews distributed awareness flyers to the traveling public at these airports as part of a 'day of action', while the ITF released its 'Zero Air Rage' Charter for Action Against Disruptive Passenger Behaviour on July 6, sending a message to governments to:

> ...have in place, or in passage, by the end of 2002, laws which give their police forces and courts the power to prosecute all incidents which occur on any flight from any country, which lands in their territory. In addition, an international convention (or an amendment to an existing Convention) under the auspices of ICAO, should be ready for all governments to sign by

the end of 2003. Governments to have in place mandatory regulations for operational training and equipment by the end of 2002.

The U.K. government fully supports the ITF's for an international treaty,[66] and so does Canada.

The ITF also calls on the air transport industry, for an integrated strategy for the prevention, management and policing of incidents.[67] In the widespread media coverage, ITF is reported to have noted a significant increase of incidents from 1,132 in 1994 to 5, 416 in 1997 world-wide. No figures were released for 1998 and 1999. These statistics underline the current subjectivity of the various methodologies for data gathering and analysis used by the organizations.

Summary

The aviation system together with the legal community, focus on a reactionary and punitive strategy as a primary tool to discourage passenger misconduct. The severity of some of the high-profile cases supports the need for this approach.

The prevention of passenger misconduct in a rapidly growing industry is not based on these reactionary measures alone. Prevention and its success depend on airline management to create a synergistic strategy in achieving marketing and operational goals. Used wisely, service and product design will help airline workers to better perform their role in maintaining social peace onboard. Understanding the impact of organizational issues on the service provider and the customer is a prerequisite for these improvements to occur, a will for responsible action a necessity.

The two risk models presented at the beginning of this chapter offer a common methodology for passenger risk assessment and management. The first model addresses an organization's predisposing traits, preconditions and mitigating factors enabling or discouraging disruptive acts.

The second illustrates the interactive nature of four key risk elements offering airlines the opportunity to address system and individual risk factors. The examples in this section show the impact of some of these factors in an operational service environment.

The section on enforcement discussed aspects of the legal safety net and stressed important issues linked to an improvement strategy. Enforcement must be structured for effective response and follow-up. As the examples reveal, this is not the case. Issues of jurisdiction need to be sorted out in the long run.

Enforcement units at the airports must monitor disruptive passenger incidents on the ground and intervene effectively. As shown in the sample statistics provided by the Airport Division of the Peel Regional Police at Lester B. Pearson International Airport in Toronto, greater efforts of vigilance are leading to increased response to passenger disturbances in the terminal and keep the problem off the aircraft. The airport community and the airline staff in particular, benefit from working closer together.

Individual initiatives as in the example of Captain Sheehy, and Sergeant Bow on behalf of the Peel Regional Police, facilitate understanding of the roles and responsibilities of crews in the specialized aircraft environment. These efforts can play a significant part in facilitating communication with enforcement leading to more effective cooperation and reliability of response. Clearly, they are not the only group of the airport community who are responsible to do so.

Successful prosecution insures a linkage to an overall system strategy for the prevention and deterrence of disruptive acts onboard. In countries where workable legislation exists, such successes are making a positive impact.

In the United States, the public believes that the government and the airlines can do more to protect the crew against attacks of passengers who loose control. Growing consumer advocacy, especially prevalent in the United States, and discussed in detail in Chapter 3, targets airline service and the government for improving the aviation system to cope with an unprecedented growth forecast over the next five years.

The Future of 'Air Rage'

The emergence of 'air rage' worldwide is linked to the complex realities of the aviation system, national and international legal, cultural and social issues. Countries that are currently lagging in addressing issues of workplace violence in all sectors of employment are not likely to deal with the hidden nature of the problem, especially, when national, organizational, and cultural factors prevent the reporting of such incidents. It will be a long way to assessing the scope of the phenomenon worldwide despite the progressive efforts by organizations such as ICAO, IATA, crew associations, and emerging research.

In countries where effective legislation allows for the prosecution of serious offenders, and with regulatory authorities and airlines continuing to address the benefit of additional regulations, e.g. training, I expect a leveling-off in the number of reported incidents over a 2 to 3 year period.

Figures released by the FAA indicate a downward trend. In 1997 the number of incidents categorized as violations of federal regulations were 320 compared to 266 in 2000. The Association of Professional Flight Attendants, representing 22,500 American cabin crews, also reported a decrease based on their own statistics. ALPA contests these figures.[68] The continued problem of a reliable database is not only linked to different methodologies but also to aviation politics.

By contrast, countries who do not have workable legislation in place, do not subscribe to preventive measures, will see a sharp rise in reported incidents as workers become more aware of their rights and look for guidance from their more industrialized international colleagues. Social change is slow and will be reflected in on-board incident patterns.

The safety of the crew and its passengers is the main concern in any of the steps that have been taken so far. As we have seen, unfortunately it is the tragic and very costly incident that rallies the will to do something. Already even the most sporadic data available clearly indicates the risks of pilot error when distracted from their primary flying duties.

The nature of regulatory authorities is to react to incidents and devise preventive measures after the fact. The airlines on the other hand have the opportunity to control the occurrence of the majority of these incidents through responsible passenger risk management aimed at prevention. In the deliberations of the Transport Canada Working Group On Prohibition Against Interference With Crew Members, Transport Canada recognized workplace violence as a foreseeable hazard.[69]

The complexity and the interactive nature of factors involved in the outbreak of 'air rage' clearly indicate that this is an issue that will have to be managed by the aviation system with the same attention as other safety and security issues. Passenger misconduct, for whatever reason, will continue to occur.

The impact of the more serious incidents discloses the rift between a technological driven system in pursuit of profits and the human side. Technology is not the main driver to achieve safety, especially when dealing with large groups of strangers in a confined environment. Similar to improved technical performance of cars resulting in a loss of safe road practices, the slackening vigilance and situation awareness is threatening the maintenance of social peace onboard. Progress is inevitable and comes with a price. It requires all parties within the aviation system to examine their values, and reconsider what service means to the safety of its workers and customers.

There are no simple solutions. 'Air rage' is a multidimensional phenomenon. Although personal inability to deal with stress, frustration and alcohol or drug abuse and mental health disorders are not a license to verbally and physically abuse others, it nevertheless sends a powerful message: The aviation system no longer serves its traveling customers well. It fails in taking timely responsibility for the growing gaps in its ability to deliver a satisfactory service; it permits victimization.

The various sectors have actively participated in creating a situation that has now become intolerable for a large number of its users. The result is an often adversarial relationship with its customers and its workers, the opposite of what is needed to produce a cooperate relationship, to maximize the safety of all. The central question is how the system will blend technology with the principles of good service, how it will develop a clear and integrated safety-service strategy and organize itself accordingly.

Collective righteousness by the aviation system or its workers can inspire much antagonism. The best service organizations take total responsibility for the customer experience (Berry, 1987), treating its customers and workers as friends. To change an entire system takes a collective will and access to resources.

Some progress has been made so far. The people customers remember most, are the cabin crew. Airlines who embrace principles of excellent customer service will plan for the events of system failures. Valuing their workers interacting with the public and protecting them from harm is part of it.

Fatalities

So far, there have been two fatalities reported as a result of 'air rage'. The first case involved a man from Finland on a Hungarian airline traveling from Bangkok to Budapest in 1998. Some passengers apparently witnessed the man taking a pill before he became violent and allegedly tried to choke a member of the cabin crew and punched a member of the flight crew. He was subsequently subdued, and a doctor administered a tranquilizer. The flight made an unscheduled landing in Istanbul because of the deteriorating condition of the passenger. He later died in hospital.[70]

The second case involved a 19-year old male passenger on Southwest Airlines 11 August 2000. The young man, traveling from Las Vegas to Salt Lake City, suddenly tried to break into the cockpit approximately 20 minutes prior to landing and was restrained with the assistance of fellow passengers. He reportedly died of an apparent heart attack after being taken into custody at Salt Lake City International Airport.[71]

The autopsy report released by the young man's family however classified his death a homicide because it resulted from 'intentional actions by another individual or individuals.' The U.S. Attorney's office stated it would not file criminal charges against passengers involved in restraining the man; an act of self-defense by frightened passengers caused this tragic death.[72] Some passengers criticized the cabin crew for lack of skills in defusing the situation, and a lack of timely intervention.

These incidents were no doubt deeply upsetting to all involved however the perpetrator was ultimately a victim himself. Acts of apparent aggression and violence may be caused by medical or drug induced condition. Cabin crews are currently left to their own devices to detect these symptoms and act in a preventive manner. Clearly, education is needed to help cabin crews in developing better diagnostic skills leading to preventive measures rather than being faced with extreme reactive measures.

The death of a perpetrator highlights issues of 'safe' restraining methods in the passenger cabin, passenger involvement, crew responsibility, and a host of legal issues. No doubt these will take center stage in the continuing debate on combating 'air rage'.

Are the Airlines Doing Enough?

Violence of any kind involves three major components: the perpetrator and the victim, society and the legal system. The legal system can punish the perpetrator and protect the victim; it has the power to reform behaviour. We see it in laws affecting drinking and driving; non-smoking regulations; we see it now addressing interference with crew members; we see the positive effects of enforcement enacting the law.

Ground is being broken in the awareness among the police, airport authorities, the courts, prosecutors and defence counsels. These achievements are due to the dedicated efforts of professional in the aviation system. No doubt years will pass before the global community has followed in the steps of the countries leading in combating air rage by closing legal loopholes.

The media is doing its part in galvanizing the public at large. Does the pronouncement of a zero tolerance policy equate with commitment? How is commitment translated into action?

In looking beyond one's own immediate responsibilities is key. Air rage is everyone's business. Getting to know the impact one's decisions have on the customer and the workers interacting with them, is part of it:

- A Chief Executive Officer who supports and tolerates socially irresponsible marketing proposal and decisions, is not committed to a zero tolerance policy. Visions of dance floors, gyms, massage facilities, nurseries, shops excite marketing mavericks, the question must be addressed: do these design features enhance a safe social onboard systems or not?
- An airline exposed to such competitive pressures has a formidable task to keep things in perspective. The bottom line is, if it jeopardizes the commitment to a zero tolerance policy, don't do it.
- A Chief Financial Officer who seeks to improve the bottom line by cutting non-regulated conflict management training, is not committed to a zero tolerance policy.
- A manager tolerating or engaging in bullying subordinates, is not committed to a zero tolerance policy; any person in a position of power within the organization or on the aircraft demeaning or bullying or intimidating their colleagues, is not committed to a zero tolerance policy.
- An employee witnessing such events and remains silent is also lacking commitment. Individuals failing to demonstrate their commitment through action enable forms of abuse and violence.

The above examples are meant to galvanize airline professionals into examining their level of commitment, to model non-violent behaviour at every step. They expect it of their customers; they need to exemplify what it means.

It starts at the top. CEO's must take on the awesome task of challenging attitudes, their own included, and scrutinize decision-making to ensure they pass the test for assessing social safety risks, especially onboard.

And what to do about the customer; is everyone worth keeping at all cost? Airlines and other service industries have started to change their minds about the cliché 'the customer is always right'. Businesses have the option to discontinue service to someone whose demands are utterly unreasonable, is abusive or threatens its employees. If such action is taken, it sends a very powerful message to the customer and a very positive one to the employees. Already some airlines have quietly implemented administrative sanctions against some passengers as an alternative to legal action.

Then there is the question of consumer education. Civil aviation authorities have taken a more active role in this endeavor as in the case of Transport Canada. A report requested by Transport Canada and submitted in 1998, also addresses the issue of printed safety information.[73]

The majority of Canadians surveyed rated the information in the seat pockets (aircraft emergency procedures cards) on the airplanes very (39 percent) or somewhat useful (45 percent). The report further determined that no more than 2 in 5 passengers read this information thoroughly.

Of greater significance is the response given to questions on other types of safety information: 75 percent of the respondents indicated to be very/somewhat interested in a brochure outlining frequently asked cabin safety questions and answers; 60 percent would be very/somewhat interested in more video taped footage of cabin safety measures; 55 percent would be very/somewhat interested in the safety statistics of individual air operators; 39 percent would be very/somewhat interested in taking a training course on cabin safety.

Revealing lesser interest by those respondents who had traveled by air the previous year, is an interesting observation and may be linked to the wide spread fear of flying, I mentioned earlier.

The media, associations and initiatives such as through the Skyrage organization are vital to maintain the focus. Airlines on the other hand prefer to let their associations speak for them, and then only in response to some public pressure.

Commitment to a zero tolerance policy includes educating its customers and the public. A wide range of broader topics and educational distribution opportunities are often missed. One of the examples is related to group sales. South African Airways developed a preventive strategy with an emphasis on timely communication with the group prior to a flight. It involves a team approach between the Group Sales Department, the Tour Operator, the group and the cabin crew. Since then, incidents of group misconduct have been significantly reduced.

What More Needs to be Done?

Ethical Management

Air travelers are strongly voicing their dissatisfaction with the airline industry. They feel their rightful expectations are ignored or insufficiently met. They feel victimized.

Ethics deal with the 'abstract ideals of fairness, justice and due process.' Applying ethics to airline management involves 'corporate social responsibility inherent in an organization's stakeholder relationships' (Lebber, 1997).[74] Although many airlines publicize their values to their employees and customers, these proclamations have little or no effect on the decisions and actions of executives. What is needed is a company ethics program that applies the values of integrity, truth and fairness to the decisions of those making critical decisions affecting others, including social peace on-board.

Too much of the focus on passenger misconduct has been on regulation, punishment and education of the passenger. The airline industry has supported these approaches because they require the least effort and cost. The real issue is the lack of a service vision as a means to safety. Governments must address the impact of an overburdened aviation system unable to cope with the high demand for air travel. As a part of national and international transportation systems governments and industry must find a better balance between the various transportation sectors. When travel experts advise passengers to ship their luggage by alternate means to their destination, it is clear that reliability of air travel, the promise of speed and convenience is no longer in its grasp. As airlines fail their customers, opportunities for other modes of transportation increase. Such is the case in Europe where high-speed trains offer a good alternative to intercity transportation.

High cost and a consistent move towards making air travel as uncomfortable and unattractive as possible, has led in good part to the current frustration. The passenger is treated as the main problem to a now inefficient system. From the time he attempts to contact a ticket agent and is put through a nightmare of the undermanned phone system, to his boarding ordeal, where he is treated as a potential terrorist and is herded into interminable lines, the passenger is quickly made aware that he is the least important element in the journey.

Efforts have to transcend mere matters of punishment. The aim of the aviation community must be more ambitious than retribution. Not that achievements so far are minor, however extraordinary restless energy is needed to lead to the prevention of such incidents in the first place. The focus is the future relationship with the air traveler. Airlines determined to preventing 'air rage' will take steps to examine what they are doing with a new commitment to ethical and safety-smart service:

- Deliver the promised service reliably by being dependable and accurate in providing what the customer reasonably expects.
- Deliver the promised service in a tangible way, with well-maintained, clean and attractive facilities, equipment and materials.
- Deliver the promised service in a responsive way, with prompt service and a willingness to serve.
- Deliver the promised service with assurance, through knowledgeable, well-trained people who are capable of conveying trust and confidence.
- Deliver the promised service with empathy by caring individualized attention, reflecting to individual customers their concerns and needs are understood.

Risk Management

Include on-board social risk management in the corporate risk management program. The four steps of common risk analysis apply:

- Hazard identification;
- Risk assessment;
- Determining the significance of a risk;
- Communicating risk information.

A good start is a well thought out database to record incidents of passenger misconduct. Superficial categorization is not helpful to managers responsible for the various factors linked to situations and conditions leading to greater risks. Poor methodology and analysis result in misinformation, easily manipulated by various interest groups with the potential for major negative implications on consumer confidence, litigation, and labour unrest.

Hazard identification includes a closer look at the types of operation, flights and routes, as well as cultural and social aspects of regions serviced. Similarly, situations conducive to passenger misconduct, the physical environment of aircraft cabins, and service features should be examined for risk potential.

How the removal of one hazard, as in the case of no-smoking regulation, has introduced another, is measurable in the number of incidents related to smoking. Airline operations benefited from this regulation by being able to reduce maintenance cost previously associated with the removal of nicotine build-up and the cleaning of ashtrays. These savings are significant, especially over the lifetime of an aircraft. Savings went directly to the corporate coffers without any portion thereof being channeled to assist air travelers to cope with their dependency, thus increasing the risk of increased passenger misconduct.

Workers' sex, age, tenure, skills and attitude are important aspects to track, especially for training, selection and developing support. The employer needs to know who are the employees most at risk in order to protect them from further victimization.

Every source of information such as complaint statistics, employee and consumer feedback are helpful in identifying hazards.

Risk assessment in the case of passenger misconduct should include employee and consumer perceptions of risks. Weighted against social, environmental and political values, the process of a cost-benefit analysis is a difficult one.

Determining the significance of risk deals with the question of what risks are acceptable. Since major airlines have already subscribed to a zero tolerance, the most logical approach would be risk aversion, however unlikely to be favoured over other approaches such as risk-balancing, risk-benefits and cost-benefit analysis.

Communicating risk information with a clear knowledge of employee and consumer perceptions of risk is essential in the educational process that will take place within an airline, the industry and the public. Providing accurate and honest communication develops trust, something sadly amiss in airline-passenger relationships today.

Simply Breathing

Reliability is the result of 'blending server and customer performances, equipment, materials, and facilities.'[75] Let us look at one example, cabin air quality. Currently, there are no cabin air quality standards. It is the aircraft manufactures in cooperation with the airlines that determine the level of customer comfort in the aircraft cabin and indirectly impact on the well-being of its occupants. It is an issue that cabin crew associations have been successful in bringing to the media's attention, raising public concern.

Cabin crew and passenger complaint and incident reports offer anecdotal evidence that the cabin environment fails in providing minimum comfort, if not causing symptoms of un-wellness. To attribute un-wellness to the cabin environment alone would be dangerous, and overlook many other important aspects outside the airlines' control.

One such aspect is the increasing age of the traveling public. More passengers with age related ailments and illnesses are using air travel as a convenient form of transportation. Acute illnesses may occur for the first time during a flight, however it does not mean the cabin environment is to blame. Long distance travel involving multiple time zones and climate changes places the body under stress, regardless of age (von Muelmann, 1995).

Feelings of discomfort affect performances, and have already been associated with passenger misconduct. When passenger discomfort issues prompt hybrid businesses to make viable profits (such as the oxygen bars at some of the airports), it clearly signals that aircraft manufacturers and airlines would benefit from greater transparency leading to less suspicion and clarification of a complex issue.

Shifting concerns raise new issues for the aviation system as perceived hazards have shifted to technology and its impact on consumer and workers health. As pressure is applied on researchers, manufacturers and airlines to explain the 'unknown', corporate values and governance at the core of decision making are scrutinized. Allegations of skewed values, deception, and unethical or illegal managerial conduct are seeds for a growing crisis in public confidence (Lebbinger, 1997).

Airlines and manufacturers need to recognize their vulnerability, and remove ambiguity through honest communication, using believable sources. These issues have now been raised to a political level in Europe and in the U.K. as previously mentioned in this chapter.

Jammed into Narrow Spaces

High-density seating is another example; abolish it. Airlines are likely taking a risk with their passengers' health. Evidence to that effect is mounting. In March 2001, medical officials from 16 major airlines and international experts addressing the link between deep vein thrombosis (DVT) met in Geneva under the auspices of the World Health Organization.

Some airlines have already made impressive investments to improve their seat configurations in economy class. Measuring complaints related to seating and passenger misconduct incidents should show improvements.

Responding to Irregular Situations

Passenger complaints point to persistently poor communication, especially when things go wrong. Communication competency standards should be defined and address:

- Apology;
- Factual and candid information regarding cause of delay, etc.;
- Anticipated time of delay;
- Impact on passengers;
- Cabin crew fact finding aimed at uncovering any special passenger needs;
- Cabin crew safety-service role during delay;
- Onboard service;
- Impact of delay on cabin environment: air conditioning etc.;
- Frequency and timing of announcements;
- Action planned for recovery.

In the midst of a problem or conflict situation, caring workers treating passengers with courtesy and respect, get high marks. Resolving problems quickly and effectively conveys trust and confidence, so does staying calm under considerable stress. Passengers love the airline representative who champions their cause and persists in finding solutions.

Train workers interacting with the public well, and take every opportunity to provide on the job coaching. Model desired communication in the workplace at all times. Apart from existing guidelines, focus on communication under duress:

- Responding in a helpful manner;
- Uncovering important and relevant information;
- Active listening;
- Empathizing with passenger's situation;
- Agreeing to disagree;
- Negotiating agreement or compromise;
- Closure.

A charming poem recalled by a retired cabin crew captures the meaning of listening versus talking:

It's silly I should talk so much,
It's funny I should choose
To keep repeating things I know,
Instead of hearing others' views.

Empathy and Assurance

Fear of flying – acknowledge it. Make it part of the training for workers dealing with the public. Include it in Human Factors training. Focus on detecting symptoms and coping methods by providing reassuring information, breathing techniques, and simple relaxation exercises.

As research has shown, aerophobia is a wide spread condition afflicting a large number of passengers. According to the Boeing study (1980), one out of four people in the U.S. do not fly because of their fear, and one in three who do get on a plane suffer greatly.

Selecting information on the mechanics of flight and safety issues from the rich body of available research can help not only the fearful flyer but also those who currently choose different modes of transportation. Arostegui, Pla, Rubio and Lorenzo (1997) argue in favour of this approach: 'We are convinced that all this information creates a sense of security and confidence in an activity which is still under a veil of mystery, arousing fears and at the same time casting a spell on people.'

Airline workers are frequently accused of being uncaring, rude and arrogant. Often, when feedback identifies these perceptions to the workers, their reaction is shock. In order to be helpful in overcoming these perceptions, the following should be discussed:

- What the passenger was looking for during the transaction: a specific service promised by the airline, information, clean and functioning facilities, individual attention, stress and fear diffusion, a special helpful act;
- What was said and how;
- Physical stance and facial expressions;
- Tone of voice;
- Alternative approaches.

Enmeshed in their own emotions, workers can easily fail to effectively communicate with the customer. Faces readily reveal the emotional nakedness. Personal stress strategies are helpful in dealing with the public.

Naturally, personal biases affect the quality of interaction, and become very apparent to the customer. Addressing the effects of personal biases in training enhances awareness and leads to better relations with the 'forgotten' passengers. Typically, passengers who do not fit the expectations of the 'perfect customer' are frequently treated condescendingly, if not worse.

Cooperative Relationships

Creating cooperative relationships takes extra effort. The air transportation system however is failing to decrease stress for its users. Research points to stress and multiple service failure as a major factor in triggering passenger misconduct.

In this book, I have addressed many of the risks under the control or influence of the system and its management. Driven by airline policies and procedures designed to maximize alienation, passengers' adaptive abilities are challenged. A Human Factors approach as part of the on-board safety equation is essential. Although there is a growing understanding of the many factors resulting in passenger misconduct, efforts to remove some of the stressors will have to be ongoing.

There is no quick fix, and punishment or reinforced cockpit doors alone will not solve the problem. Cumulative stress stemming from poor service and lack of employee skills contribute to triggering passenger misconduct and are management's responsibility. Service design must be integrated with safety objectives.

The more serious cases, unpredictable due to predominant personal risk factors associated with the perpetrator, are more difficult to prevent. Expertise from the medical community, psychiatry, and security professionals is needed to help in handling these situations more effectively. A coordinated and broad-based, multi-faceted strategy with cooperation from all parties involved is the foundation for improvements.

Notes

[1] J. Reason coined the term 'latent and active failures' in his research surrounding human error in complex systems.

[2] Corporate culture and safety is the focus of research affecting different modes of transportation as supported by a selection of researchers (Ginett, 1997; Hackman, 1997, 1987, 1984; Marske, 1991, 1997; McDonald and Fuller, Meshkati, 1997,1994, 1988, 1989b,; Pidgen, 1991; Pidgeon and O'Leary, 1994; Reason, 1997; Suggs, 1997; Westrum, 1999, 1997).

[3] Working Group On Prohibition Against Interference With Crew Members – Report, 20 June 2000, pp. 12.

[4] *Avflash,* 14 February 2000, vol. 6, 07a.

[5] Fear of flying is the general term used for phobic reactions to all aspects surrounding flying. Phobic reactions are characterized by excessive and groundless fears of external objects and conditions.

[6] Scharfetter (1991) distinguishes phobia from fear in that the person experiencing a phobia attack clearly understands that there is no substantiation of real danger. Specific situations and objects trigger phobia.

[7] Acrophobia or fear of heights.

[8] Claustrophobia or fear of enclosed spaces.

[9] Agoraphobia or the fear of public or open spaces.

[10] Byrne-Crangle, M. (1995), 'Fear of flying: an investigation into aerophobia and its treatment', in R. Fuller, N. Johnston and N. McDonald (eds), *Human Factors in Aviation Operations,* Proceedings of the 21st Conference of the European Association for Aviation Psychology, vol. 3, pp. 170.

[11] Some of the airlines who conduct or have conducted seminars for passengers suffering from fear of flying are: Aerolineas Argentinas, Air Canada, Air France, American Airlines, British Airlines, Iberia, Lufthansa, Swissair, and United Airlines.

[12] *AJN,* 11 August 2000.

[13] Cabin crew associations are not supporting an outright ban, however do express the need for more responsible service policies.

[14] The Indianapolis Star, 22 April 2000.

[15] Drugs fall into two main categories, those used to treat medical conditions and those used for mind alteration. For air travelers it is important to consult with their physician prior to planning a flight. This is especially important prior to long-haul trips, since side effects of a drug or its irregular use due to time zone changes could be considerably altered. Human Factors for aviation professionals deal extensively with this issue.

[16] *AJN,* 23 March 2000.

[17] *Kuam News,* 17 May 2000.

[18] *San Francisco Chronicle,* 21 September 2000.

[19] Stress, its various stages, and coping with stress have been researched extensively. Human Factors publications regularly feature the topic and its significance for flight crew performance. Generally, three types of stress are referred to: physical, physiological and emotional. Physical stressors include environmental conditions; physiological stressors refer to the effects of fatigue, fitness, nutrition, sleep loss, as some of the examples on the body. Emotional stressors include social and emotional factors (Trollip and Jensen,1991).

[20] Dahlberg, A. (1999), 'Passenger Risk Management' in Proceedings of the Ninth International Symposium on Aviation Psychology, Columbus, OH.

[21] Lazarus and Folkman (1998) link the accumulation of smaller hassles to stress intensification and its deleterious effects.

[22] Two cases were reported in August 2000, one on an All Nippon flight from Tokyo to San Francisco, the other on a United Airlines flight from Narita, Japan to San Francisco. Both man have been charged.

[23] ICVS results indicate female victims of sexual harassment are between 16 and 34 years of age. (See: *Violence at Work*, Geneva: International Labour Organization, pp. 27-28.)

[24] Chappell and Di Martino report wide spread sexual harassment in the airline industry. According to one study conducted in Australia in 1989, more than one-half of the cabin crews surveyed claimed to be so harassed. (See: *Violence at Work*, Geneva: International Labour Organization, pp. 70.) Interestingly, sexual harassment was not an issue raised by CUPE in the deliberations of the Transport Canada Working Group.

[25] Relative humidity ranges between 10 and 25 percent Davis, 1994), with the level of humidity decreasing on long-haul flights at altitudes of 37,000 to 39,000 feet.

[26] This view is also supported by researchers Arostegui, Pla, Rubio and Lorenzo (1997).

[27] Air Canada is one of the operators having developed and implemented such administrative sanctions.

[28] The effects of inflight violence are the topic of a paper presented by Michael Sheffer, husband of Renee Sheffer (founders of SKYRAGE) in a paper presented at the *Tenth International Symposium on Aviation Psychology*, Columbus (1999).

[29] *CBC News*, 14 January 2000.

[30] *Air Line Pilot*, April 1999, pp.10-13, 54.

[31] *Air Line Pilot*, April 1999, pp.13.

[32] BLUE LINE, Canada's National Law Enforcement Magazine, February 2000: 'Airport Policing in the new millennium' by Sgt. Malcolm Bow.

[33] *Air Line Pilot*, September 2000, pp. 20.

[34] Associated Press, July 6 2000.

[35] *Airwise News*, 28 January 1999.

[36] Aviation security international, February 1999, pp. 24-27.

[37] *Calgary Herald*, 12 December 2000.

[38] *Calgary Herald*, 8 January 2000.

[39] *Calgary Herald*, 20 April 2000.

[40] *Globe & Mail*, 11 December 1999.

[41] *CBC News*, 18 November 1999.

[42] *Calgary Herald*, 7 August 1999.

[43] *Reuters*, 31 January 2001.

[44] IATA Inflight Services web site.

[45] Las Vegas Review Journal, 22 January 2000.

[46] *The Times*, 14 December 1999.

[47] *LA Times*, 23 November 1999.

[48] *Washington Post*, 23 October 1999.

[49] uk.news.yahoo.co, 991028.

[50] News & City, this is London.co.uk, 15 October 1999.

[51] *Telegraph*, 21 August 1999.

[52] *Calgary Herald*, 22 July 1999.

[53] Electronic Telegraph, 7 July 1999.

[54] Electronic Telegraph, 29 May 1999.

[55] Electronic Telegraph, 11 March 1999.

[56] Calgary Herald, 26 July 2000.

[57] *CNN News*, 17 July 2000.

[58] *CNN News*, 18 July 2000.

[59] The Associated Press, 20 July 2000.

[60] *Chapter 3*, The Air Traveler – A Human Factors Issue.

[61] *Travel News*, 20 July 2000.

[62] *Reuters*, 26 July 2000.

[63] The Seattle Times, 30 July 2000.

[64] *Commission of the European Communities,* 'Communication from the Commission to the European Parliament and the Council – Protection of Air Passengers in the European Union', COM(2000) 365 final, Brussels, 21 June 2000.

[65] One Japanese carrier reported incidents more than doubled over the period of 1998 to 1999. The leading categories were abusive and unruly behaviour rating at 37 percent, followed by smoking violations and alcohol abuse at 21 percent, and sexual offences at over 7 percent.

[66] International Transport Workers' Federation – media information, July 2000.

[67] International Transport Workers Federation, "Zero Air Rage" Charter, July 2000.

[68] *Air Line Pilot,* September 2000, pp. 19.

[69] Working Group On Prohibition Against Interference With Crew Members – Report, 20 June 2000. p. 15.

[70] BBC News World Europe, 5 December 1998.

[71] *AJN,* 14 August 2000.

[72] *Air Crash Rescue News,* 17 September 2000 & *Airwise News,* 18 September 2000. The autopsy apparently revealed that the young man was strangled. Bruises and scratches were found on his torso, face and neck.

[73] *The Goldfarb Report 1998,* p. 3.

[74] Lebber, O. (1997), *The Crisis Manager,* Lawrence Erlbaum Associates, Inc., Mahwah, pp. 314.

[75] Berry, L.L. (1995) 'on Great Service – A Framework for Action', The Free Press, p. 87.

Appendix A – Denial of Carriage

The following example of passenger misconduct incident, and a decision rendered by the Canadian Transportation Agency is interesting for a number of reasons:

- Passenger's perception of service entitlement;
- Passenger's reaction to authority;
- Passenger's coping strategies in the on-board environment;
- Crew members' conflict resolution skills;
- The impact of a prior incident with the same passenger on the captain's decision to refuse carriage on a subsequent flight;
- The risk of early intervention as a stressor.

DECISION NO. 492-A-1998[1]

FACTS: Mr. M. had purchased a round trip ticket from Mirabel to Acapulco. Prior to the departure of Flight No. TS365 from Acapulco to Mirabel, Mr. M. was ordered by the flight captain to deplane the aircraft for safety reasons. Mr. M. was stranded and was only able to return to Mirabel the following day on an American Airlines, Inc. flight.

POSITIONS OF THE PARTIES: Mr. M. indicates that he and the flight captain on duty, Mr. R.P., had previously had an encounter which had resulted in an argument following the arrival of carrier X's Flight No.TS364 in Acapulco from Mirabel on January 17, 1998. In this respect, Mr. M. stresses the fact that he frequently travels on business, especially with carrier X. He was therefore in a position to compare the various services and, when Flight No. TS364 had arrived in Acapulco on January17, he openly voiced his dissatisfaction with respect to the poor quality of the services he had received on board that particular flight. Captain P. had not appreciated Mr. M.'s comments and an argument had ensued between the two. Mr. M. states that on the return flight, he took his seat on board the aircraft without any problem. Following that, Captain P. saw him and advised him that he was hesitant to have him on board as he considered him to be a 'potential safety risk to his passengers' given the aggressiveness he had shown upon the arrival of Flight No. TS364. In Mr. M.'s view, there was no reason for this to occur. A few minutes later, and without prior notice, Mr. M. was asked to deplane following the Captain's orders. As a result of this refusal to carry him on board Flight No. TS365, Mr. M. had to book a seat on another flight to return to Mirabel, without any assistance or concern on the part of carrier X. Mr. M. holds carrier X and Captain P. responsible for the prejudice which he suffered and demands the following: an apology from both Captain P. and carrier X; the assurance on the part of carrier X that similar incidents will not recur; monetary compensation for costs and losses incurred, including the loss of one day's pay, the inconveniences and the fatigue he incurred due to the fact that his travel plans were changed. He also requests compensation for damages (such as humiliation, defamation, etc.) which he suffered. For its part, carrier X explains that Captain P. had been asked by the flight director to deal with Mr. M. on board Flight No. TS364. According to her, the passenger was unjustifiably rude and disagreeable with the flight attendants. Mr. M. went so far as to throw things (blankets) at them and used coarse language. After the arrival of Flight No. TS364 in Acapulco, carrier X was able to use only one set of stairs for deplaning passengers given the number of aircraft at the airport at that time. These stairs had been directed toward exit L-2 of the aircraft. Frustrated at not being among the first passengers to deplane, Mr. M. started yelling, again using coarse language. Captain P. attempted to explain the situation to Mr. M., to no avail. In light of the previous incidents which occurred on Flight No. TS364 in Acapulco on January 17, Captain P. attempted to talk to Mr. M. prior to the departure of Flight No. TS365 on March 7, 1998, to ensure that he would not endanger the safety of the other passengers and of his crew. At that time, Mr. M. lost his temper and demonstrated the same aggressive behaviour as on Flight No. TS364. Captain P. therefore decided to have Mr. M. removed from the aircraft. Carrier X submits that the refusal to carry Mr. M. was justified given the very difficult circumstances of the preceding Flight No. TS364 on January 17, Mr. M.'s behaviour, attitude and language prior to the take-off of Flight No. TS365 on March 7, 1998, and given Captain P.'s well-founded fear for the safety of the other passengers and his crew if he had allowed Mr. M. to remain on board. Without any admission of responsibility whatsoever, carrier X offered to reimburse Mr. M. for his return flight with American Airlines, Inc., provided he submit the appropriate proof, as well as an undetermined, yet reasonable, amount for the lost wages for one day.

OTHER SUBMISSIONS: On July 22, 1998, carrier X provided the Agency with a copy of various flight reports which relate the incidents described herein. Carrier X, however, requested that the above-mentioned letter and the documents attached thereto be kept confidential by the Agency. By Decision No. LET-A-248-1998, the Agency provided its determination that the release of the information would not cause specific direct harm to carrier X, and therefore denied the application of carrier X. The documents were copied to Mr. M. for comment. In response, Mr. M. indicates that his behaviour, his attitude and his language when boarding the aircraft on March 7, 1998 were not at issue, as confirmed in his view by carrier X's flight reports. Further, nothing indicates that there was a serious or well-founded fear with respect to the safety of the passengers on board Flight No. TS365. According to him, Captain P.'s decision seems to have been for the most part motivated by the incident which occurred on January 17 upon arrival of Flight No. TS364. Mr. M. therefore submits that his treatment by carrier X was abusive and that the carrier has not complied appropriately with the terms and conditions contained in its tariff.

ANALYSIS AND FINDINGS: The Agency is the Canadian aeronautical authority responsible for the economic regulation of air carriers. Among its responsibilities, it must follow up on consumer complaints by ensuring that the provisions of the

Canada Transportation Act, S.C., 1996, c. 10, as well as those of its attendant regulations, including the *Air Transportation Regulations*, SOR/88-58, as amended (hereinafter the ATR), are complied with by air service operators. The Agency must also ensure that air carriers abide by the terms and conditions of their respective tariffs. Pursuant to section 110 of the ATR, air carriers who provide an international air service to and from Canada must file their tariffs with the Agency. Pursuant to paragraph 122(*c*) of the ATR, tariffs shall contain the terms and conditions of carriage clearly stating the air carrier's policy in respect of, amongst other matters, the refusal to transport passengers or goods. The Agency has examined carrier X's tariff applicable to the service described herein and notes that rule 12.1(A) thereof stipulates that the carrier may refuse to carry a passenger for safety reasons. After having fully examined the evidence provided by the parties, as well as all documents pertaining to the complaint, including the air carrier's tariff, the Agency determines that carrier X has not contravened the terms and conditions of its tariff. The Agency is of the view, based on the information on file, that Captain P.'s reasons for refusing to transport the passenger were justified and in compliance with rule 12.1(A) of the air carrier's tariff. The Agency notes that Transport Canada has established a working group to develop measures and strategies for dealing with aggressive passengers. The Agency will therefore provide a copy of this Decision to Transport Canada to be submitted to the working group.

CONCLUSION: In light of the above findings, the Agency has determined that carrier X has not contravened the terms and conditions of its tariff, and the applicable regulatory requirements. Consequently, the complaint filed by Mr. M. is hereby dismissed.

Note

[1] *File No. M4370/A328/98.*

Appendix B – Carrier Compliance

DECISION NO. 269-C-A-2000[1]

PRELIMINARY MATTER: On December 8, 1999, carrier X requested that the in-flight reports that were submitted as part of its reply on December 6, 1999, be maintained confidential by the Agency. By Decision No. LET-A-21-2000 dated January 26, 2000, the Agency determined that the release of the information would not cause specific direct harm to carrier X, and therefore denied the application by carrier X. The documents were copied to Mr. A. for comment.

ISSUE: The issue to be addressed is whether carrier X has complied with the *Air Transportation Regulations*, SOR/88-58, as amended (hereinafter the ATR) and the CTA in respect of this matter.

FACTS: Mr. A. and a companion purchased round-trip economy class tickets for travel between Montréal and Toronto. Mr. A's carry-on baggage included a travelling suit bag. After boarding Flight No. CP1611, a discussion ensued between in-flight personnel and Mr. A. about the stowing of his suit bag. Following discussions and further to the aircraft captain's orders, Mr. A. was informed that he had to leave the plane. After retrieving his belongings, Mr. A. and his companion were escorted off the plane. Carrier Z provided Mr. A. and his companion with new tickets for Flight No. CP1613, departing Montréal one hour later than their originally scheduled flight.

POSITIONS OF THE PARTIES: Mr. A. submits that, upon boarding the aircraft, he asked an in-flight agent if she could place his carry-on suit bag in the closet located in the first/business class section of the aircraft. The in-flight agent refused to oblige, and informed Mr. A. that he would have to place his bag in the overhead compartment himself. Mr. A. states that he proceeded to his seat and his companion placed their carry-on bags in the overhead compartment. He kept his suit bag on his lap, but was informed by the in-flight agent that he could not keep his bag on his lap, and that he would have to place it in the overhead compartment. Mr. A. states that he requested assistance from the agent to stow his bag in the overhead compartment, explaining that he suffered from back problems, but the agent refused. Mr. A. submits that a second in-flight agent informed him that it was Mr. A.'s responsibility to place his suit bag in the overhead compartment. According to Mr. A., the in-flight agent stated 'Would you like me to call security to take you off the flight?', to which Mr. A. replied that if he was going to be taken off the flight, he wanted his money back. Shortly after, Mr. A. submits that a security guard approached him, and after a discussion, the security guard placed the suit bag in the overhead compartment. Mr. A. states that a customer service representative then informed him that he had to leave the plane, as per the aircraft captain's orders. He and his companion were escorted off the plane. According to Mr. A., he was the victim of ill treatment by badly trained and abusive employees, which caused him severe anxiety, embarrassment and delay in returning to his daily professional commitments. For its part, carrier X submits that its staff acted in accordance with the provisions of its tariff, as well as Transport Canada directives in respect of a passenger who does not comply with instructions from a flight crew member. In response, Mr. A. submits that he did comply with the instructions he was given by the flight crew members to the limit of his physical capacity. Mr. A. further states that only after he had been properly assisted in his efforts by the security guard to comply, and compliance with the flight crew's directive was achieved, did the flight crew members seek the intervention of the aircraft captain to have him removed from the plane.

OTHER SUBMISSIONS: The in-flight reports submitted by carrier X indicate that despite several requests from the in-flight crew, Mr. A. did not comply with the instructions to stow his bag, as per Transport Canada regulations. One of the reports states that although Mr. A. indicated that he had a sore back, he did not preboard, ask for assistance, or have an attendant. Furthermore, the reports indicate that several passengers thanked in-flight crew for removing Mr. A. because of his obnoxious and rude behaviour. Mr. A. submits that when he was denied access to the closet and ordered to his seat, he complied. When he was asked to place his suit bag in the overhead compartment, he did not refuse to comply but informed the flight attendants he could not do it himself because he is physically unable to lift weight above his shoulders.

ANALYSIS AND FINDINGS: The Agency is the legislative authority responsible for the economic regulation of air carriers. The role of the Agency is to ensure that the provisions of the CTA, as well as those of its attendant regulations, including the ATR, are complied with by air service operators. Section 55 of the CTA defines a tariff as a schedule of fares, rates, charges and terms and conditions of carriage applicable to the provision of an air service and other incidental services. Subsection 67(1) of the CTA provides, among other matters, that the holder of a domestic licence shall publish and make available for public inspection its tariffs for the domestic service it offers. Pursuant to paragraph 107(1)(*n*) of the ATR, tariffs shall contain the terms and conditions of carriage clearly stating the air carrier's policy in respect of, among other matters, the refusal to transport passengers or goods. The Agency has examined carrier X's tariff applicable to the service described herein. The terms and conditions of carriage applicable to carrier Z's flight from Montréal to Toronto on April 5, 1999 were governed by the air carrier's Domestic General Rules Tariff in effect at the time of travel. Paragraph 6 of Rule 34CP, Refusal to Transport, of carrier X's Domestic General Rules Tariff provides, in part, that:

> (6) Carrier may refuse to transport or may remove at any point any passenger whose behavior is interfering or has interfered with the safety or comfort of any other passenger or any crew member. Passenger shall discontinue any such behavior immediately upon request of a crew member.

After having fully examined the evidence provided by the parties, as well as all finds that carrier X's domestic tariff clearly states the carrier's policy in respect of refusal to transport passengers or goods. The Agency is also of the opinion that carrier X did not act in a manner inconsistent with the above-quoted tariff provision.

CONCLUSION: In light of the above findings, the Agency has determined that carrier Z has not contravened paragraph 107(1)(*n*) of the ATR. The Agency is of the further opinion that carrier X did not depart from its tariff in respect of this matter. Consequently, the complaint filed by Mr. A. is hereby dismissed.

Note

[1] *File No. 4370/C14/99.*

Appendix C – FAA Guidance

The following documents serve as references to guidance material by the FAA that played an important part in shaping the formal process for other regulatory authorities, industry and crew associations in addressing passenger misconduct.

The FAA produces the first document of importance in the form of an advisory circular:

10/18/96   120-65   AFS-200

INTERFERENCE WITH CREWMEMBERS IN THE PERFORMANCE OF THEIR DUTIES

1. PURPOSE. This advisory circular (AC) provides information to air carriers, crewmembers, law enforcement officers, and the general public regarding methods which may be used to manage and reduce the instances of passenger interference with crewmembers. This AC provides general information about the types of subjects which could be included in an operator's program. In addition, examples of this type of information are provided in Appendices 1, 2, 3, 4, 5, and 6. These examples are based on material provided by the airline industry. Airlines wishing to adapt these samples for their own use should carefully read the legal disclaimer at the top of each sample.

2. RELATED CFR SECTIONS. Title 14 of the Code of Federal Regulations (14 CFR) sections 91.11, 108.10, 108.11, 108.19, 121.317, 121.575, 135.121, and 135.127.

3. DISCUSSION. It is important that both the traveling public and crewmembers have a safe environment when on board an aircraft. Pertinent regulation says that no person may assault, threaten, intimidate, or interfere with a crewmember in the performance of the crewmember's duties aboard an aircraft. The majority of passenger violations are filed under this rule. Additional regulations prohibit the boarding of passengers or serving alcohol to passengers who appear to be intoxicated. Passengers must also obey passenger information signs such as the no smoking and seatbelt sign. In addition, they must obey the instructions of the crewmembers regarding compliance with these signs.

a. Crewmembers, airlines, and Federal Aviation Administration (FAA) personnel have concerns about the increase and nature of occurrences where passengers intimidate, threaten, and/or interfere with crewmembers. In addition, passengers have complained to the FAA and the airlines about being intimidated and uncomfortable because of some of the actions of fellow passengers. Therefore, this document has been prepared to provide guidance about the type of programs that are designed to reduce the number of problems and the stress caused by these incidents.

b. In order to properly discuss this matter, it is necessary to make some attempt to define the types of occurrences. For purposes of this AC, the FAA has divided the types of events into broad categories which are contained in the chart in Appendix 1. This chart provides one means of categorizing passenger misconduct. Additional examples of defining passenger misconduct are contained in the various appendices to this document. Any of these examples is acceptable. In addition, these are not the only means of categorizing passenger misconduct; an air carrier can develop its own methods of defining these occurrences.

4. <u>POLICY OF THE OPERATOR</u>. One of the most important aspects of any program dedicated to the reduction of violence in the workplace is the commitment of each individual, including those with management responsibilities. Therefore, partnerships which include employees with differing responsibilities, and appropriate government personnel should be formed to develop procedures, handle violence, and provide assistance to individuals who are involved in passenger disturbances.

a. Airlines should make it clear to all employees what actions should be taken when an incident occurs that meets any of the broad categories found in Appendix 1. The operator's program should involve all personnel who have direct contact with passengers. The emphasis of the program should be on keeping dangerous passengers off the airplanes. There should be clear lines of responsibility regarding the handling of these events. These responsibilities should include offering and/or providing counseling for those who are involved in or who witness the events. Employee assistance groups can also play an important role in providing this assistance.

b. Operators should establish policies which define the operator's philosophy concerning zero tolerance. Appendices 2, 3, 4, and 5 contain programs which have worked well and provide examples of policy statements where air carriers have provided information about their zero tolerance philosophies.

c. It is important that the operator provide the public with the appropriate information and thereby provide a safe environment for crewmembers and for the traveling public. The operator should provide material to passengers regarding the seriousness of inappropriate behavior on an airplane, including failure to follow instructions from crewmembers. Further, the material should contain information to the passengers about the consequences of their actions including possible fines and incarceration. Public awareness information can be in the form of pamphlets passed out at airport gates, included in ticket envelopes, articles in onboard magazines, posters in gate areas, public address announcements, information given in video tapes, or any other method that management believes will convey the message to the public. A sample of the information that could be disseminated is included in Appendix 6.

5. <u>WRITTEN PROGRAMS</u>. Operators should make it clear to all employees what actions should be taken when an incident occurs and involves a crewmember. This program should be included in crewmember, security personnel, and other appropriate manuals. The written information should be disseminated to all employees of the air carrier who could have the responsibility for handling a situation with a dangerous passenger. A sample form carried on board the flights by crewmembers giving information about one method of handling onboard incidents is provided in Appendix 5.

a. It is important that written programs be developed with employees who are familiar with the security aspect of the airline, including crewmembers. These are the people who have the most experience with and are familiar with the local law enforcement jurisdictions and will be the most likely to help educate their staff about passenger interference with crewmembers.

b. The written program should encourage employees to promptly report cases of interference on reporting forms such as the sample provided in Appendix 5. The written report should contain at least the names of the crewmembers, the date, flight number, seat number, origin/destination of flight, the name, address, and description of the offending passenger, and the names and addresses of witnesses. If positive identification is not established by the crewmembers, then the written program should provide guidance on securing identification through appropriate airline personnel, law enforcement, or other methods as appropriate.

c. The written program should also provide information about personnel in the company who should contact law enforcement and the FAA. Information should also be provided regarding how crewmembers may directly contact the FAA and law enforcement on their own.

d. In addition, the written program should include information regarding filing complaints against passengers. The process of pursuing violations requires an ongoing commitment and should not be taken lightly. The employee may be required to testify in any subsequent court proceedings.

6. TRAINING. Air carriers should provide training for crewmembers and other responsible personnel for handling passengers who interfere in the performance of crewmember duties. The training should acknowledge that it is not desirable to have cockpit crewmembers leave their stations, especially in cases where there are two cockpit crewmembers. Nevertheless, the training should also acknowledge the authority of the captain and that the decision to leave the cockpit is the responsibility of the captain. Airlines may want to include training on passenger misconduct in the required training during crew resource management, hijacking, and other unusual situations. Regardless of how the training is provided, it should include information which will help the crewmember recognize those situations which may, when combined with traits of some passengers, create stress. The training should include information about how to manage conflict situations, such as:

a. Responding to Imminent Danger. If the passenger becomes abusive, solicit help from other crewmembers, other employees, or passengers to help restrain the individual. Usually the other person will be another flight attendant; however, at times it may be wise to involve passengers. This is especially true when the flight is operating with one flight attendant. Cockpit crewmembers should be kept well-informed. The decision to have a crewmember leave the cockpit is the responsibility of the captain. Flight attendants should provide as much information as possible to the cockpit crew. The captain should be given the passenger's name (if possible), description and the name and description of traveling companions, seat number, and if medical attention is needed. Inform the captain if you wish authorities to meet the inbound flight.

b. Reporting the Information. Flight attendants should be informed on the use of the forms which the air carrier has developed for the purpose of handling passengers who cause disturbances. When law enforcement officials are called to meet the flight, crewmembers need to be informed that written statements will be taken upon arrival and that they may be called to testify in court.

7. LAW ENFORCEMENT AND FAA RESPONSE. Incidents of interference with crewmembers could be a serious violation of regulations and may warrant a response from local law enforcement or the Federal Bureau of Investigation (FBI). In most cases, the initial response will be provided by the airport law enforcement department or, if there is no resident law enforcement

unit at the airport, the department having overall responsibility for law enforcement support to the airport.

a.   When the incident of interference is sufficient to warrant a response from law enforcement, the captain should notify dispatch/flight following and request a law enforcement representative and an air carrier representative meet the airplane upon arrival at the gate.

b.   Law enforcement response may involve interviewing one or more members of the crew, other passengers who witnessed the incident, and the subject passenger(s). Action may be taken by the law enforcement department responding or a report may be forwarded by the local law enforcement to the FBI and FAA. In some cases, the FBI and FAA may be called to meet the arriving airplane. This will usually happen for the more serious incidents such as assault, intimidation using a dangerous weapon, threat or actual attempted sabotage or hijacking.

c.   It should be noted that every incident of interference will not warrant a response from law enforcement personnel. A crewmember must ask a law enforcement representative to meet the aircraft. In order to take action, there must be a legal basis for an officer to do so. For example, physically assaulting a crewmember would warrant law enforcement action. However, for example, if the incident involves failure to fasten a seatbelt, there may not be a legal basis for criminal action from the local law enforcement unit. This does not imply that a formal complaint needs to be filed by a member of the crew for action to be taken. If there is a serious incident, the action may be taken by the government for violation of a criminal statute or for violation of specific regulations.

d. The airline should inform all members of the crew that full cooperation is necessary in reporting an incident in a timely manner and providing statements if requested by local law enforcement, the FBI, or the FAA.

e. Jurisdiction and authority for action is a consideration for any law enforcement officer's response and an arrest may not be the result in every case. Any action taken must be within the scope of authority for the law enforcement officer. If an arrest is made by the airport law enforcement unit or, if a case is referred to the FBI for investigation, prosecution still rests with the appropriate office of the prosecuting attorney.

f. Reports forwarded to the FAA may result in joint investigative efforts by the FAA and FBI. Cases where the FBI declines to investigate may still be worked by the FAA and could result in a civil penalty for the passenger involved in interfering with the crewmember.

g. The FAA has asked its principal inspectors and managers to emphasize review of incidents involving interference with crewmembers. A partnership effort between the FAA, FBI, local law enforcement, and the industry which emphasizes communication and cooperation should lessen the number of incidents of interference.

The same procedures should be followed for international flights and the law enforcement response will be those of the destination government.

Thomas C. Accardi

This sample airline information should be reviewed by each airline's legal department to assure that it accurately states the airline's policies and the legal duties, responsibilities, and rights of the airline and airline personnel. The FAA does not provide legal advice about the specifics of tort and criminal law.

APPENDIX 1.

CATEGORY ONE Flight attendant requests passenger to comply. (These are actions which do not interfere with cabin or flight safety such as minor verbal abuse.)	Passenger complies with request.	There is no further action required by the flight attendant. (Such an incident need not be reported to the cockpit, the carrier, or the FAA.)
CATEGORY TWO Flight attendant requests passenger to comply.	Passenger continues disturbance which interferes with cabin safety such as continuation of verbal abuse or continuing refusal to comply with federal regulations (such as failure to fasten seatbelt when sign is illuminated, operation of unauthorized electronic equipment). In addition, the crewmember should follow company procedures regarding cockpit notification.	After attempting to defuse the situation, the captain and the flight attendant will coordinate on the issuance of the Airline Passenger In-flight Disturbance Report or other appropriate actions. The flight attendant completes the report. Completed report is given to appropriate company personnel upon arrival. In turn, company personnel may file the incident report with the FAA.
CATEGORY THREE	Examples: (1) when crewmember duties are disrupted due to continuing interference, (2) when a passenger or crewmember is injured or subjected to a credible threat of injury, (3) when an unscheduled landing is made and/or restraints such as handcuffs are used, and (4) if operator has program for written notification and passenger continues disturbance after receiving written notification.	Advise cockpit, identify passenger, then cockpit requests the appropriate law enforcement office to meet the flight upon its arrival.

This sample airline information should be reviewed by each airline's legal department to assure that it accurately states the airline's policies and the legal duties, responsibilities, and rights of the airline and airline personnel. The FAA does not provide legal advice about the specifics of tort and criminal law.

APPENDIX 2. SAMPLE AIRLINE POLICY BULLETIN -- ASSAULTS ON EMPLOYEES

AC 120-65 10/18/96

Although rare, assaults on employees by customers do occur. In many cases, the assault occurs simultaneously with other actions which interfere with the duties of a crewmember. Until today, the Company has handled these cases on an individual basis with the employees involved. This policy is being adopted in an effort to help employees at all levels better understand their rights and responsibilities in the event of an assault.

In many jurisdictions, an assault is defined as an action taken toward an individual that creates a threat of bodily harm, or the apprehension of physical injury. In some jurisdictions, abusive or suggestive language, unless used in a manner that creates the threat of violence or harm, is not considered an assault. If physical contact occurs, the incident is usually defined as battery. Often, an event involving an assault or battery is generally referred to as an assault.

SPECIAL PROTECTION FOR CREWMEMBERS

Crew interference is governed by federal regulation (Title 14 of the Code of Federal Regulations (14 CFR) section 91.11). Crew interference is defined as an incident where a passenger assaults, threatens, intimidates or interferes with a crewmember while in performance of crew duties on board an aircraft. THIS AIRLINE WILL NOT TOLERATE ASSAULT, THREATS, INTIMIDATION, AND INTERFERENCE. ANY EMPLOYEE WHO IS SUBJECTED TO ASSAULT WHILE AT WORK WILL RECEIVE COMPANY SUPPORT (INCLUDING LEGAL ADVICE...PAID ABSENCE TO APPEAR IN COURT DURING A CRIMINAL PROCEEDING).

The decision to press charges requires an ongoing commitment by the employee and should not be taken lightly. The employee may file a complaint or be required to testify in any subsequent court proceedings.

The Company will provide legal counsel and supervisory assistance in pursuing appropriate action to any employee who is subjected to abuse, physical violence, or intimidation on the job.

APPENDIX 2. SAMPLE AIRLINE POLICY BULLETIN -- ASSAULTS ON
EMPLOYEES (Cont'd)

An employee may also pursue a civil action against a party who has committed an assault or battery. A civil action is brought for the purpose of recovering money damages.

In addition, support is available to any employee who is the victim of an assault through the airline's Employee Assistance Program at (phone number). The Company, jointly with the union, also provides a critical incident stress debriefing team which is available to flight attendants in certain circumstances.

It is important that employees report assaults immediately to the Company. All reports will receive follow-up by the appropriate department. All reports of crew interference are filed with the FAA for recording and possible civil enforcement action. Additional reports obtained for the FBI or local police are attached to the crew reports to assist the FAA in their investigation and assignment of appropriate penalty.

It is important to obtain as much information about the offender as possible. A name and address, as well as witness statements, are valuable. At a minimum, a description of the attacker, including physical characteristics, will be important when pursuing legal action. In an aircraft situation, the passenger's assigned seat designation often allows the Company to obtain information through its reservations' records.

As always, employees are expected to be understanding in trying to resolve the frustrations of our customers. However, no one can be expected to tolerate physical abuse of any kind.

This sample airline information should be reviewed by each airline's legal department to assure that it accurately states the airline's policies and the legal duties, responsibilities, and rights of the airline and airline personnel. The FAA does not provide legal advice about the specifics of tort and criminal law.

APPENDIX 3. SAMPLE AIRLINE INFORMATION BULLETIN

AIRLINE SECURITY INFORMATION BULLETIN

TO: ALL PUBLIC-CONTACT, FLIGHT OPERATIONS, IN-FLIGHT, AND ASSOCIATED MANAGEMENT PERSONNEL

SUBJECT: ASSAULTS ON EMPLOYEES

Just as any of us would take action if a family member needed help, each of us may feel an obligation to respond when a fellow employee needs help. Our corporate values clearly support this by asking us to show respect for each other as individuals and demonstrating integrity in everything we do.

When a fellow employee is in distress, for any reason, we should immediately and effectively assist that person. This certainly applies in cases of assault. Not coming to the aid of an employee in distress as a result of a customer's actions constitutes a clear failure to adhere to our corporate values. If serious physical assaults are ignored by pilots or managers, for example, basic safety and security may be compromised---and an individual's dignity violated. When a coworker or crewmember ignores an assaulted employee, that employee most likely will feel ignored and abandoned by the airline as well.

It is very important to be aware that authorities should be called for assistance with unruly customers or instances of out-and-out battery. Furthermore, flight officers have an obligation to follow-up on an assault which occurs on an aircraft by requesting that authorities meet the trip and by filing a "Captain's Report of Crewmember Interference."

The following are questions and answers which will provide you with more information about the issue of assault in the workplace.

Q. What does "ASSAULT" actually mean?

A. Many jurisdictions define assault as an action taken toward an individual

APPENDIX 3. SAMPLE AIRLINE INFORMATION BULLETIN
(Cont'd)

that creates a threat of bodily harm or the apprehension of physical injury. Abusive or suggestive language, if it is not utilized in a manner that creates the threat of violence or harm, is not considered an assault in some jurisdictions. Generally speaking, if physical contact should occur, the incident is defined as battery. Often, any event involving an assault or battery is referred to as an assault.

Q. What is the company's policy regarding assault?

A. At (<u>airline</u>), assault will not be tolerated. Any employee who is subjected to assault while at work will receive company support, including legal assistance and paid absence to appear in court during a related criminal proceeding.

Q. Will the company provide me with a lawyer?

A. The company will provide legal assistance and supervisory assistance in pursuing appropriate criminal remedial action to any employee who is subjected to abuse, physical violence, or intimidation on the job. The airline, however, will provide legal advice throughout the proceedings.

Q. What if I want to file a civil suit?

A. The decision to pursue a civil action against a party who has committed an assault or battery belongs to the employee. A civil action is brought for the purpose of recovering monetary damages. The company will, however, support the employee, counsel him or her as to their rights, and even assist in finding or retaining an attorney.

Assault in the workplace is a very serious issue. By lending a helping hand when necessary, we can support each other.

APPENDIX 4. SAMPLE PROCEDURES DEALING WITH FLIGHT ATTENDANT ASSAULT (Cont'd)

POLICY: Title 14 of the Code of Federal Regulations (14 CFR) section 91.11 states, "No person may assault, threaten, intimidate, or interfere with a crewmember in the performance of the crewmembers' duties aboard an aircraft being operated."

PROCEDURES:

During Boarding, at the Gate, or Taxi-Out:
- If the boarding flight attendant or agent identifies a passenger exhibiting inappropriate behavior, they should confer and prior to the passenger boarding, notify the captain and the lead agent. An example of inappropriate behavior could be a passenger who appears to be intoxicated, or has questionable medical problems that could be an immediate threat to other customers or themselves.
- If the passenger is on board the aircraft, the lead flight attendant will notify the captain of the passenger's name, seat number, and the nature of the problem.
- Reports of this nature can be reported during the sterile cockpit period if necessary.

After Takeoff/En Route:
- The captain will be notified by the lead flight attendant if any passenger displays disruptive behavior, appears to be intoxicated, or is smoking on a nonsmoking flight.
- After attempting to defuse the situation, the captain and the lead flight attendant will coordinate on the issuance of the Airline Passenger In-flight Disturbance Report to the passenger.
- It may not be safe for a cockpit crewmember to leave the cockpit. If the passenger becomes abusive, solicit help from other cabin crewmembers, other company employees, or passengers to help restrain the individual.
- Upon arrival, the captain will make a Public Address System Announcement (PA) requesting all passengers remain seated.
- The lead flight attendant will coordinate with the captain to identify passengers involved to the authorities.

Postflight:
- All flight attendants will complete a flight attendant report. Verify the name

and address, if possible, of the passenger engaging in misconduct, and of any witnesses.

- Flight attendants need to be prepared to make a verbal and written statement to the local authorities upon landing. Flight attendants will retain a copy of any written report.
- The captain will facilitate any meetings with local authorities and/or appropriate airline personnel.
- Followup assistance, such as legal counseling, medical assistance, or personnel counseling will be provided by the flight attendant department or other appropriate departments.

This sample airline information should be reviewed by each airline's legal department to assure that it accurately states the airline's policies and the legal duties, responsibilities, and rights of the airline and airline personnel. The FAA does not provide legal advice about the specifics of tort and criminal law.

APPENDIX 5. SAMPLE REPORTING FORM AIRLINE PASSENGER IN-FLIGHT DISTURBANCE REPORT

Date: _____

Flight #: _____   Departure City:_____   Arrival City: _____

Passenger Information: Name: _____

Seat #: _____

Address: _____

_____

_____

Description of Incident: _____

_____

_____

_____

_____

_____

Witness Name: _____   Seat #: _____

Address: _____

_____

Phone #: _____

F/A Name: _____

Employee #: _____   Base: _____

F/A Signature: _____

Captain Name: _____

Employee #: _____   Base: _____

Captain Signature: _____

This sample airline information should be reviewed by each airline's legal department to assure that it accurately states the airline's policies and the legal duties, responsibilities, and rights of the airline and airline personnel. The FAA does not provide legal advice about the specifics of tort and criminal law.

APPENDIX 5. SAMPLE REPORTING FORM (Cont'd)

NOTICE: Your behavior may be in violation of Federal law.
You should immediately cease if you wish to avoid prosecution and your removal from this aircraft at the next point of arrival.

This is a formal warning that Federal law prohibits the following (reference Title 14 of the Code of Federal Regulations (14 CFR) parts 91 and 121):

- Threatening, intimidating, or interfering with a crewmember (section 91.11)
- Smoking on a nonsmoking flight or in the lavatory (section 121.317)
- Drinking any alcoholic beverages not served by a crewmember or creating an alcohol-related disturbance (section 121.575)

An incident report will be filed with the FAA. If you do not refrain from these activities you will be prosecuted. The Federal Aviation Act provides for civil monetary fines and, in some cases, imprisonment.

APPENDIX 6. SAMPLE PASSENGER DISTURBANCE ADVICE FORM

TITLE 14 OF THE CODE OF FEDERAL REGULATIONS (14 CFR) SECTION 91.11

Please be advised that interference with crewmembers' (including flight attendants) duties is a violation of Federal law.

An incident report may be filed with the Federal Aviation Administration regarding a passenger's behavior.

Under Federal law, no person may assault, threaten, intimidate, or interfere with crewmembers (including flight attendants) in the performance of their duties aboard an aircraft under operation.

Federal law permits penalties for crew interference to include substantial fines, imprisonment, or both.

Appendix D – CTA Decision

Canada Transportation Agency Decisions

The reader may ask, why so much emphasis on the disabled? The answer is four-fold:

1. To underline the effects of increasing legislation on airline operations and its challenges;
2. To address the question, why persons with disability do not appear to be afflicted by air rage;
3. To examine how a person with disability deals with substantial service failures related to air travel using the legal safety net;
4. Lessons to be learned.

Because of physical restrictions, the use of physical threat is not an option. By necessity and not by choice, methods for solving problems are honed differently. The following examples speak for themselves. If not for the support of working legislation, societal attitudes towards the disabled would remain unchallenged, and these situations would very likely remain unresolved. Legislation provides the most effective framework for channeling service failures. The lessons for the air traveler without disability are:

1. Educate yourself. Know your rights and obligations that come with the purchase of the ticket. It is a contract resulting in a conditional relationship. Regulatory safety requirements and company rules define not only the airline's but also your role and responsibilities.
2. Be informed what consumer agencies monitor airline performance. They provide clear guidelines how to file a complaint and ensure follow-up.
3. Examine you anger strategies. Make use of the formal processes available to all consumers. Channel your anger productively.
4. Stay calm, no matter what. Shouting, blaming, and otherwise unchecked expressions of frustration to anger obstruct problem solving. This type of conduct does not make allies
5. Develop patience.

The following case was widely covered by the Canadian media:[1]

DECISION NO. 475-AT-A-2000[2]
ISSUE: The issue to be addressed is whether carrier X's refusal to allow Mrs. B. to

travel unattended constituted an undue obstacle to her mobility and, if so, what corrective measures should be taken.

FACTS: Mrs. B. is 88 years old and has limited mobility due to arthritis. Consequently, she uses either a walker or a cane.

Plans had been made for Mrs. B. to travel with carrier X on February 15, 2000 from the Montréal International Airport - Dorval, Quebec, to Washington, D.C. to visit her daughter and her son-in-law, Dr. I.J. K. From there, she was to travel with them on February 17, 2000 to Los Angeles, California.

At the time of reservation, wheelchair assistance had been requested for Mrs. B. to facilitate boarding. She arrived at the airport accompanied by her son and her live-in companion, then proceeded to check in and was escorted through U.S. Customs. Subsequently, she was preboarded using a boarding chair and taken to her seat at the front of the aircraft.

In an effort to facilitate access to the onboard washroom facilities, the flight attendant offered Mrs. B. a seat at the back of the cabin. Mrs. B. declined and the flight attendant determined that Mrs. B. was non-ambulatory, non-self-reliant, and could not travel unattended. Despite her protests and insistence that she could travel alone and that family members were expecting her in Washington, the attendant maintained his position and insisted that she deplane.

Mrs. B. was taken to the Hilton Hotel where she remained with an airline attendant until arrangements could be made for someone to accompany her to Washington. On February 16, 2000, Dr. I.J. K. travelled from Washington to Montréal in order to accompany Mrs. B. on her trip to Washington. He purchased a round-trip ticket, at a cost of $534.20 US, and had to cancel all his appointments for the day.

On June 30, 2000, carrier X issued a letter to Ms. H. enclosing reimbursement for the airline tickets of Mrs. B. and Dr. K. in the amounts of $327.85 US and $534.20 US, respectively.

POSITIONS OF THE PARTIES: Ms. H. states that her grandmother's disability is minor and that she can walk, although not long distances. Because of her condition, she has learned to adapt when travelling, including refraining from using the washroom facilities onboard aircraft for periods of up to four hours. On longer flights, she requires assistance to move from her seat to the washroom but does not require assistance to use the facilities. In addition, she always books an aisle seat in the forward part of the aircraft as this is her preference.

In Ms. H.'s opinion, there was no reason to reassign a seat at the back of the cabin as, according to Canadian legislation, air carriers are required to provide assistance to passengers with disabilities to move to and from an aircraft washroom, other than by carrying them.

In addition, she submits that when her father, Dr. K., spoke with the carrier X supervisor regarding the requirement for an attendant, safety was raised as an issue, but not Mrs. B.'s ability to use the washroom. According to the flight attendant on duty, she would have to deplane if she did not have an attendant. He explained that because of the weather conditions, Mrs. B. would be unable to leave the aircraft in case of an emergency. According to Ms. H., the flight attendant claimed that this was a carrier X policy.

Despite Mrs. B.'s reassurance that she indeed could travel alone, as she had previously done on a number of occasions, and notwithstanding her objection, Ms. H. submits that her grandmother was 'rudely and roughly' deplaned by the flight attendant and another man. They physically removed her from her seat onto an aisle chair, then left her on the tarmac where she remained for several minutes before being taken into the airport terminal. Ms. H. submits that Mrs. B. was freezing, but was mostly shocked, humiliated, angry and depressed. Furthermore, during the transfer from the aircraft seat to the onboard wheelchair, Ms. H. submits that her grandmother hit her head which led to swelling.

During her overnight stay at the hotel, Mrs. B. began to feel the effects of having been left in the snow and cold at the airport. Ms. H. submits that, as a result, Mrs. B. had to consult a doctor during her stay in Los Angeles, California, and had to take prescription medicine for an eye infection.

Ms. H. suggests that if an attendant was indeed required as claimed by the flight attendant, options other than deplaning Mrs. B. were available. For example, airline personnel could have asked one of the passengers to volunteer to provide assistance in case of an emergency. If, in carrier's X opinion, Mrs. B. could not travel alone, then she should not initially have been allowed to board the aircraft and someone should have contacted her family members. Ms. H. submits that the safety concerns that were raised because only one flight attendant was on duty were unfounded and were not an excuse to deplane Mrs. B.

Ms. H. is of the view that carrier X has failed to ensure that an appropriate level of training has been given to all of its employees who provide services and may be required to interact with the public or make a decision regarding their transportation.

Ms. H. submits that her grandmother was a victim of discrimination and that the flight attendant made his determination not to allow her to travel based on her appearance. As a result of the incident described herein, Mrs. B. is now afraid to travel non-stop between Montréal and Washington.

Ms. H. requests that carrier X provide her with a written confirmation that disciplinary action has been taken against the airline employee involved, and that the carrier amend its policy on the transportation of passengers with disabilities who travel alone. Further, she requests that carrier X reimburse the costs of Mrs. B.'s visit to the doctor and the related prescription; Dr. K.'s air fare and his consulting and travel fees; costs related to the filing of the complaint; monetary compensation; goodwill compensation, as well as denied boarding compensation.

Ms. H. is also of the opinion that, where a jetway is not available to access aircraft, a lift should be provided to assist in boarding and deplaning passengers with disabilities.

Carrier X's code-share partner, carrier Y, did not provide any explanation for the incident except to extend its apologies and state that the matter had been discussed with carrier X representatives.

Carrier X submits that Mrs. B. declined the offer to occupy a seat at the back of the aircraft cabin, indicating that she was unable to use the washroom on her own. Consequently, the flight attendant discussed this situation with the captain and they both referred to carrier X's Safety and Emergency Procedures Manual, more specifically to the definition of 'non-ambulatory, non-self-reliant' which states that these passengers are unable to move within the aircraft cabin without assistance, are incapable of self-care and, therefore, must be accompanied. Consequently, based on Mrs. B.'s statement and on the fact that airline personnel do not provide assistance to passengers inside the washroom, carrier X personnel concluded that she was non-ambulatory, non self-reliant, and that she had to be accompanied during the flight to Washington.

While the Manual also provides a definition of passengers who are non-ambulatory, yet self-reliant, meaning that these passengers require assistance to move about the aircraft cabin, but are independent and do not require special services or attention, carrier X submits that the flight attendant did not overlook this paragraph of the manual, but rather determined that Mrs. B. was incapable of self-care based on her determination that she was unable to use the washroom by herself.

However, carrier X does acknowledge that its personnel did not apply the policy correctly as they overlooked the paragraph of the manual pertaining to the acceptance of a person's self-determination that he or she will not require extraordinary services during a flight and will not need an attendant. It extends its deepest apologies for the inconvenience which resulted from this regrettable incident.

Carrier X states that it has reviewed the incident with the pilot and flight attendant in question and reminded them of its policy and, more specifically, that a passenger's self-determination that he or she will not require extraordinary services must be accepted. Carrier X trusts that these measures will prevent the recurrence of incidents similar to those experienced by Mrs. B.

Carrier X further states that Mrs. B.'s limited mobility raised safety concerns as only one attendant would be on duty during the flight. According to carrier X's procedures, when operating a flight with only one attendant, he or she is responsible for assigning a fellow passenger to assist a passenger with a disability in an emergency situation. Carrier X employees are periodically trained in this respect.

Carrier X submits that it has no information to support the assault and battery charge. Its employees have no recollection of Mrs. B. hitting her head when she was removed from the aircraft. The Customer Service Manager who was present remembers that Mrs. B. was wearing her coat when she was deplaned in a covered area, and that she was promptly taken inside the terminal and not left in the snow on the tarmac.

ANALYSIS AND FINDINGS: In making its findings, the Agency has considered all of the material submitted by the parties during the pleadings.

The Agency notes that, in its reply, carrier X raises two issues to explain why Mrs. B. was deplaned: firstly, Mrs. B.'s inability to use the onboard washroom, therefore to travel alone, and, secondly, the safety issue.

With respect to the first issue, the Agency has considered the statements made by both parties regarding the exchange that took place on February 15, 2000 between Mrs. B. and the flight attendant. According to carrier X, Mrs. B. declined to relocate to a seat at the back of the aircraft as 'she was unable to use the washroom by herself.' In Ms. H.'s words, Mrs. B. rather indicated to the flight attendant that 'she would not have to go to the rest room.'

Based on the contradicting statements of both parties, the Agency is of the opinion that there was clearly a misunderstanding between the flight attendant and

Mrs. B. as to whether or not she required assistance. The Agency is of the opinion that this misunderstanding resulted from the lack of communication between the flight attendant and Mrs. B. In light of Mrs. B.'s objection, the flight attendant should have had a thorough discussion with her in order to determine what her needs were and why she preferred to remain in her assigned seat at the front of the aircraft. In doing so, the flight attendant would have realized that Mrs. B. was not unable to use the washroom on her own, but rather elected not to use it at all. This lack of communication and information led the flight attendant and the flight captain to wrongly determine that Mrs. B. was 'non-ambulatory, non-self-reliant.' Based on the definition of non-ambulatory passengers provided in carrier X's Safety and Emergency Procedures Manual, they determined that Mrs. B. could not travel alone. The Agency is of the opinion that the definition of non-ambulatory self-reliant passengers was overlooked, and more importantly, as acknowledged by carrier X, so too was the carrier's policy to accept the determination made by a passenger or his or her representative that he or she will not require extraordinary services during a flight and that an attendant is not required.

The Agency is concerned that the actions of carrier X's personnel were in contradiction with this policy and that this oversight on the part of its personnel had serious consequences for those involved. As a result of being deplaned, Mrs. B. experienced a delay in her travel plans she had to remain in a hotel with a stranger and suffered the stress of not knowing the outcome of the situation which could impact on Dr. K.'s business travel plans. Further, the incident resulted in Dr. K. having to interrupt his activities and incur expenses in order to take a flight to Montréal to act as an attendant for Mrs. B. on the flight to Washington.

The Agency notes carrier X's safety concerns given that only one flight attendant would be on duty during the flight. The Agency finds, however, that deplaning was not the only option available to Air Canada personnel, if indeed they were convinced that Mrs. B.'s limited mobility posed a risk. The responsible flight attendant could have assigned a fellow passenger to assist with the evacuation in case of an emergency as carrier itself stated during pleadings that this was part of its emergency procedures in such situations.

Regardless of the motive for deplaning Mrs. B., the Agency finds that the non-acceptance of Mrs. B.'s self-determination regarding the assistance, if any, required during her travel constituted an obstacle to her mobility. The obstacle was undue as it disrupted her travel plans and inconvenienced her family members. The obstacle could easily have been avoided if the flight attendant had entered into a discussion with Mrs. B. to determine and ensure that she could travel unattended and did not require extraordinary services.

The Agency notes Ms. H.'s statement that, in accordance with Canadian legislation, namely the *Air Transportation Regulations*, SOR/88-58, as amended, air carriers are required to provide assistance to passengers with disabilities to move to and from aircraft washroom facilities. While these Regulations apply to domestic flights only and not to transborder or international flights, the Agency notes that carrier X's International Passenger Rules and Fares Tariff states that 'The carrier will accept the disabled person's determination as to self-reliance.'

With respect to Ms. H.'s statement that lifts should be provided at airports that do not provide jetways for boarding and deplaning aircraft, the Agency notes that the procedure consisting of manually boarding and deplaning passengers using a Washington or straight back chair is current in the Canadian airline industry as well as in other countries. The Agency notes, however, that it is in the process of studying boarding mechanisms for small aircraft. This involves consultations with airlines and airport authorities in view of improving airport equipment and accessibility within airport terminals and aircraft.

The Agency has considered the applicant's request for compensation, i.e. costs pertaining to the filing of the complaint and subsequent pleadings, medical expenses, as well as Dr. K.'s loss of work time and round-trip ticket.

While the Agency recognizes the seriousness of the incident experienced by Mrs. B. and is concerned by the actions of carrier X's flight personnel, it does not have jurisdiction to award damages for stress and duress.

However, pursuant to subsection 172(3) of the *Canada Transportation Act*, the Agency finds that, as a result of the undue obstacle to Mrs. B. mobility, the travel expenses incurred by Dr. K. are admissible for compensation. Although carrier X reimbursed Dr. K. for the costs of the airline tickets purchased to travel from Washington to Montréal in order to accompany Mrs. B. back to Washington, the Agency is of the opinion that carrier X should also reimburse Dr. K. for all out-of-pocket expenses.

With respect to Mrs. B.'s medical expenses while in Los Angeles, the Agency finds that insufficient evidence was provided during the pleadings to allow the Agency to determine whether Mrs. B.'s condition was a direct result of the incident or whether it was latent. Consequently, the Agency finds that this claim is inadmissible.

Concerning the claim for denied boarding compensation, the Agency finds that according to carrier X's Canadian General Rules Tariff, NTA(A) 241, on file with the Agency, Mrs. B. is not entitled to denied boarding compensation.

Finally, the Agency does not consider it appropriate to award costs in the present matter.

CONCLUSION: In light of the above, the Agency finds that the non-acceptance of Mrs. B.'s self-determination that she would not require extraordinary services during the flight, which resulted in her subsequently being deplaned, constituted an undue obstacle to her mobility.

Consequently, the Agency hereby directs carrier X to:

- submit a report on the corrective measures it has taken to ensure that its employees are aware of the incident described herein to avoid any recurrence of such incidents;
- submit the training records of the flight attendant who interacted with Mrs. B. on February 15, 2000; and
- reimburse all of Dr. K.'s out-of-pocket expenses upon receipt of supporting documents.

Carrier X is required to provide the Agency, within thirty (30) days from the date of this Decision, with a written confirmation that the above measures have been taken.

Should the parties not reach an agreement with respect to the out-of-pocket expenses, they may refer the matter back to the Agency for a determination of cost and the issuance of an order to that effect.

Notes

[1] *Calgary Herald,* 21 July 2000.
[2] *File No. U3570/00-23,* 18 July 2000.

Appendix E – Interview Assessment

Cabin Crew Interview Assessment (for females)[1]

The airline and the crew member have been de-identified to honor the request of the person having kindly submitted this document.

Applicant's Name	Date	Location

FIRST IMPRESSION of her as she
gets up and walks toward you.
(Write in SAP before you change
your mind.)

APPEARANCE (If she wears glasses, make special note in 'General Remarks')

FACIAL FEATURES	Beautiful Pretty Attractive[2]		Ordinary Plain Unattractive
PROFILE	Regular		Receding chin Sloping Forehead Other
EXPRESSION	Alert	Friendly	Dull Hard Hostile Tight-lipped Apprehensive Theatrical Grimaces
SMILE	Easy Warm Frequent		Average Infrequent Unattractive
TEETH	Even White Sparkling		Crooked Overlapped Missing Dentures Yellow Discoloured Dull Protrude Wide gapped
COMPLEXION	Flawless Clear Smooth		Odd spot Pimples Rough
HAIR	Nice Natural Neat		Lifeless Untidy Unnatural

HANDS	<u>Slim</u> <u>Nice</u>	Well kept		Chubby	Reddened	Rough

FINGERS	<u>Long</u>	Slender	Short	Fat	Bony

NAILS	<u>Well manicured</u> Good shape		Too long	Bitten	Broken

FIGURE	Excellent	<u>Proportioned</u>	Average Thin Heavy-hips Over/under weight Flat chested

LEGS (Asses also shapely knees & ankles) How would legs look with short skirt?	V. shapely Delightful	Acceptable Sl. Heavy Sl. thin Good <u>Acceptable</u>	Heavy Thin Shapeless Bowed Knock-kneed Heavy thighs Thick knees Bony knees Thick ankles Poor Awful

DRESS	Smart Colourful <u>Appropriate</u>	Drab Ill-fitting Inappropriate

ACCESSORIES	<u>In good taste</u>	Unsuitable

CARRIAGE	<u>Upright</u> Graceful Lithe	Pigeon-toed Droopy Round-shouldered

STATURE	<u>Ladylike</u>	Acceptable	Slouches	Fidgets	Inelegant

VOICE (Tone)	<u>Clear</u> Strong <u>Gay</u> <u>Pleasant</u>	Quiet	Dull	Monotone Accented

SPEECH: English Language

SPOKEN (incl. Vocabulary, Pronounciation & Grammer)	Excellent <u>Good</u>	Acceptable (odd error)	Average	Fair	Poor	Accented Garbled

CONVERSATION	Easy Natural Enthusiastic	Stilted Tongue-tied Doesn't finish sentences
	Interesting Completes sentences	Condescending

PERSONALITY

MANNER	Outgoing <u>Polished</u> Warm <u>Friendly</u> Enthusiastic	Reserved Lifeless Conceited Aggressive Unfriendly Cold Brash

ATTITUDE	Tactful	<u>Flexible</u>	Calm	Sensitive Impatient	Vague Arrogant Rash

CONFIDENCE	<u>Well poised</u>	Composed	Nervous Tense Overconfident

MATURITY	Realistic	<u>Well adjusted</u>	Naïve	Unrealistic Irresponsible

HEIGHT (checked)	WEIGHT (checked)	U/C VISION (if known)

GENERAL REMARKS (include special considerations, Company recommendations, etc.)

This young lady seems to meet our appearance requirements except for her teeth. They are nice and white but also big.
Nice personality and appeared sincere. She understands reserve coverage too. Was apparently anemic and has undergone tests to that effect.[3]

DATE: 6 May 1969

(Signed)

Notes

[1] This unedited example is taken from an actual Canadian airline that used this interview assessment form designed in 1968, until the early 1970's.

[2] Ratings applicable to this applicant are underlined.

[3] Comments by interviewing Supervisor.

Bibliography

Andrews, P.W. (1997), 'The Impact of Corporate Culture on Transportation Safety', paper presented at the *NTSB Symposium on Corporate Culture and Transportation Safety.*

Augustin, S. (1999), 'Stress and It's Relationship to In-Flight Violence', in *Proceedings of the Tenth International Symposium on Aviation Psychology,* Columbus, OH.

Aviation Safety Reporting System. Search Request No. 5362 (1998, July) and No. 5599 (1999, April). *Passenger Misconduct Incidents.* NASA, ASRS Office.

Bainbridge, L. (1999), 'Processes Underlying Human Performance', in D.J. Garland, J.A. Wise, V.D. Hopkin (eds), *Handbook of Aviation Human Factors,* Lawrence Erlbaum Associates, London, pp. 153-155.

Becker, T.A. (1992), 'Passenger Perceptions and The Marketing of Airline Safety – The Awakening of Airline Safety', San Diego, California.

Beerman, B. and Meschkutat, B. (1995), 'Psychological factors at the workplace: Taking account of stress and harassment,' Federal Institute for Occupational Safety and Health, Dortmund.

Beh, H. and McLaughlin (1991), 'Mental performance of air crews following layovers on transzonal flight', *Ergonomics,* vol. 34, pp. 123-135.

Berry, L.L. (1995), *On Great Service – A Framework for Action,* The Free Press, New York, pp. 63, 65, 68, 87, 130, 147, 247n.

Bor, R. (1999), 'Unruly Passenger Behaviour and In-flight Violence: A Psychological Perspective', in *Proceedings of the Tenth International Symposium on Aviation Psychology,* Columbus, OH.

Bornas, X. and Tortella-Feliu, M. (1995), 'Description and psychometric evaluation of a self-report instrument for fear of flying assessment,' in R. Fuller, N. Jonston and N. McDonald (eds), *Human Factors in Aviation Operations,* Proceedings of the 21st Conference of the European Association for Aviation Psychology (EAAP), Vol. 3, pp. 174-179.

Bow, M. (2000), 'Airport Policing in the new millennium'. *BLUE LINE, Canada's National Law Enforcement Magazine,* February 2000, pp. 6-7, 9.

Boyd, Neil, (1993), 'Violence in the workplace in British Columbia: A Preliminary Investigation,' Simon Fraser University.

Byrne-Crangle, M. (1995), 'Fear of flying: an investigation into aerophobia and its treatment,' in R. Fuller, N. Jonston and N. McDonald (eds), *Human Factors in Aviation Operations,* Proceedings of the 21st Conference of the European Association for Aviation Psychology (EAAP), Vol. 3, pp. 169-173.

Chapell, D. and Di Martino, V. (1998), *Violence at Work,* Geneva: International Labour Organization, pp. 63, 70, 72.

Chute, R., Wiener, E.L., Dunbar, M.G. and Hoang V.R. (1995) 'Cockpit/Cabin Crew Performance: Recent Research', in *Proceedings of the 48th International Safety Seminar, Flight Safety Foundation.*

Chute, R. (1995), 'Cockpit/Cabin Communication: I. A Tale of Two Cultures',

International Journal of Aviation Psychology, 5 (3), pp. 257-276.

Cornish, M. and Lopez, S. (1994), 'Changing the workplace culture through effective harassment remedies,' in *Canadian Labour and Employment Law Journal* Scarborough, Ontario Vol. 3, No. 1.

Collingridge, D. (1992), The Management of scale: Big organizations, big decisions, big mistakes, Routledge, London.

Connell, L.J., Mellone, V.J. and Morrison, R., (2000), 'Cabin Crew Safety Information and The NASA Aviation Reporting System', in *Proceedings of the Seventeenth Annual International Aircraft Cabin Safety Symposium*, Southern California Safety Institute.

Cotton, D. (1997), 'Dealing with Disruptive Passengers: A Shared Responsibility', in Proceedings of the 14th Annual International Aircraft Cabin Safety Symposium and Technical Conference, Southern California Safety Institute, pp. 86-92.

Crowley, J.S., Wesenten, N., Kamimori, G., Devine, J., Iwanyk, E. and Balkin, T. (1992), 'Effect of high terrestrial altitude and supplement oxygen on human performance and mood', in *Aviation, Space and Environmental Medicine*, vol. 63, pp. 696-701.

Dahlberg, A. (1999), 'Passenger Risk Management' in *Proceedings of the Tenth International Symposium on Aviation Psychology*, Columbus, OH.

Dahlberg, A., (1997), 'Flyer's Rage' *INSIGHT, IATA*,Vol. 5, Issue 4, July/August, pp.34-35, 48-50.

Dahlberg, A., (1997c), 'Added Value – Customer Perceptions of Safety Culture', in *Proceedings of the Ninth International Symposium on Aviation Psychology*, Columbus, OH.

Dahlberg, A., (1997b), 'Customer Perceptions of Cabin Safety Management', *in Proceedings of the 14th Annual International Aircraft Cabin Safety Symposium and Technical Conference*, Southern California Safety Institute, pp. 346-355.

Dahlberg, A., (1997a), 'Selling Safety' *INSIGHT, IATA, vol.*5, Issue 1, January/February, pp. 10, 42-44.

Dahlberg, A., (1996), 'The Unruly Passenger - Dangerous Goods of Another Kind' *INSIGHT, IATA, vol.* 4, Issue 6, November/December, pp. 30-31, 46.

Dahlberg, A. (1995) 'Customer service quality: improving customer service through Human Factors' management in the passenger cabin' in R. Fuller, N. Johnston and N. McDonald (eds), Human Factors in Aviation Operations, Proceedings of the 21st Conference of the European Association for Aviation Psychology (EAAP), Vol. 2, pp. 270-276.

Dahlberg, A. (1995), 'Customer Expectations and Perceptions of The Quality of Safety Management' (unpublished).

Davis, J.R. (1994), 'Passenger Medical Care In-Flight', in *Proceedings of the Eleventh Annual International Aircraft Cabin Safety Symposium and Technical Conference*, Southern California Safety Institute, pp. 222-225.

Dawson D. and Reid, K. (1997), 'Equating the Performance Impairment associated with Sustained Wakefulness and Alcohol Intoxication', in *Proceedings of the 14th International Aircraft Cabin Safety Symposium*, Southern California Safety Institute, pp.288-295.

Dean, R.D. and Whitaker, K.M. (1980) 'Fear of Flying: Impact on the U.S. air travel industry', Boeing Company Document , BCS-00009-RO/Om.

Deiner, E. (1980), 'Deindividuation: The absence of self-awareness and self-regulation in group members' in P. Paulus (eds), *The psychology of group influence,* Lawrence

Erlbaum Associates, Hillsdale, New Jersey, pp. 209-242.

Denison, D.R. (1990), *Corporate culture and organizational effectiveness*, John Wiley & Sons, New York.

Donnerstein, E. and Wilson, D. (1976) Effects of noise and perceived noise on ongoing and subsequent aggressive behavior, in *Journal of Personality and Social Psychology*, vol. 34, pp. 774-781.

Douglas, M. and Wildavsky, A. (1982), 'Risk and Culture: An Essay on the Selection of Technical and Environmental Dangers', Berkeley, University of California Press.

Edwards, M. and Edwards, E. (1990), *The Aircraft Cabin: Managing the human factors*, Gower Technical, Brookfield.

Emenaker Kovarik, L., Graeber, R.C. and Mitchell, P.R. (1999), 'Human Factors Considerations in Aircraft Cabin Design', in D.J. Garland, J.A. Wise, V.D. Hopkin (eds), *Handbook of Aviation Human Factors,* Lawrence Erlbaum Associates, London, pp. 389, 391, 399-400.

Endsley, M.R. (1999), 'Situation Awareness in Aviation Systems' in D.J. Garland, J.A. Wise, V.D. Hopkin (eds), *Handbook of Aviation Human Factors*, Lawrence Erlbaum Associates, London, pp. 258-259, 265-266.

Freedman, J. (1975), *Crowding and Behaviour*, Freeman Press, San Francisco.

Ernsting, J. (1984), Mild hypoxia and the use of oxygen in flight, in *Aviation, Space and Environmental Medicine,* vol. 55, pp. 407-10.

Fuller, R. (1994), 'Behaviour analysis and aviation safety', in Johnston, N., McDonald. N., & Fuller, R. (Eds.), *Aviation Psychology in Practice.* Avebury Technical, Aldershot, pp. 173-189.

Geller, E.S. (1994), 'Ten principles for achieving a total safety culture', *Professional Safety*, pp. 18-24.

Ginnett, R.C. (1997), *'Building a Culture for Team Safety: By Design and By Default'*, paper presented at the NTSB Symposium on Corporate Culture and Transportation Safety.

Ginnett, R.C. (1993), 'Crews as groups: Their formation and their leadership', in E.L. Wiener, B.G. Kanki and R.L. Helmreich (eds), *Cockpit resource management*, Academic Press, New York.

Greenspan, E.L. (1987) *The Case for the Defence,* Macmillan of Canada, Toronto.

Guidelines for Handling Disruptive/Unruly Passengers, 1st Edition Effective 1998, IATA.

Hackman, J.R. (1997), ' Why Teams Don't Work', paper presented at the *NTSB Symposium on Corporate Culture and Transportation Safety.*

Hackman, J.R. (1993), 'Teams, leader, and Organizations: New direction for crew-oriented flight training', in E.L. Wiener, B.G. Kanki, and R.C. Helmreich (eds). *Cockpit resource management,* Academic Press, san Diego, pp. 47-69.

Hackman, J.R. (1984), 'The Transition that hasn't happened', in J.R. Kimberly and R.E. Quinn (eds), *New Futures: The challenge of managing corporate cultures,* Dow Jones-Irwin, Homewood.

Hampson, B. (1999), 'Outsourcing of Flight Crew Training', *The Journal of Professional Aviation Training*, vol. 1, Number 3, Jan-Feb, pp 14-21.

Hampden-Turner, C. (1987), 'Humanistic Psychology and the Crisis of the American Economy,' a keynote presentation made to the International Association of Humanistic Psychology at Mills College, Oakland, California.

Hayward, B. (1995) 'Organisational change: the human factor' in R. Fuller, N. Johnston and N. McDonald (eds), Human Factors in Aviation Operations,

Proceedings of the 21st Conference of the European Association for Aviation Psychology (EAAP), vol. 1, pp. 63-68.

Hibbert, A. (1997), 'Medical Implications for the Stressed Flyer', in *Proceedings of the 15th International Aircraft Cabin Safety Symposium*, Southern California Safety Institute, pp. 264-256, 265-268.

Hill, A. (1997), 'Passenger Non-Compliance: How much is too much?', in *Proceedings of the 14th International Aircraft Cabin Safety Symposium*, Southern California Safety Institute, pp. 64-73.

Horodniceanu, M. and Cantilli, E.J. (1979), *Transportation System Safety*, Lexington, MA.

ICAO Circular 240-AN/144 (1993), *'Investigation of Human Factors in Accidents and Incidents'*, Human Factors Digest No.7.

ICAO Circular 247-AN/148 (1993),*'Human Factors, Management and Organization'*, Human Factors Digest No. 10.

International Transport Workers' Federation (2000), *'Air Rage – The Prevention and Management of Disruptive PassengerBehaviour'*. ITF, Civil Aviation Section.

Inzana, C.M., Driskell, J., Salas, E. and Jonston, J. (1996) 'Effects of Preparatory Information on Enhancing Performance Under Stress', *Journal of Applied Psychology*, vol. 81, no. 4, pp. 429-435.

Jenkins, E.L. (1996), *Violence in the workplace: Risk factors and prevention strategies*, National Institute for Occupational Safety and Health (NIOSH), Publication No. 96-100, Washington, DC, US Government Printing Office.

Johnston, N. (1993), 'Managing Risk and apportioning blame', IATA 22 Technical Conference, Montreal.

Jungermann, H. and Goehlert, C. (2000), 'Emergency evacuation from double-deck aircraft' in M.P. Cottam, D.W. Harvey, R.P. Pape and J. Tait, 'Foresight and Precaution', Proceedings of ESREL 2000, SARS and SRA-Europe, A.A. Balkema/Rotterdam/Brookfield, pp. 989-992.

Karlins, M. and Cheah, C.H. (1997), 'Speak up for Safety! Encouraging Flight Attendants to Share their Safety Concerns with Pilots' in *Proceedings of the Ninth Int'l Symposium on Aviation Psychology*, Columbus, OH, pp. 532-535.

Klemmer, J. (1990), *Firing on All Cylinders*, Macmillan of Canada.

Knight, S. and Butcher, N. (1995), 'Safety and Service – Is there Conflict in the Cabin?' in *Proceedings of 12th Annual International Aircraft Cabin Safety Symposium*, Southern California Institute, pp. 26-35.

Krajc, T. and Pausch, R. (1998), 'The Perception And Significance Of Flight Safety From The Customer's Point Of View In The Deregulated Market' Diplomarbeit, Technische Universitat Darmstadt.

Lauber, J.K. (1993), 'A safety culture perspective', in Proceedings of the Flight Safety Foundation 38th Annual Corporate Aviation Safety Seminar, pp. 11-17.

Lawrence, C. (1995), 'Violence in the Workplace/Airplane Cabin', in *Proceedings of the 12th Annual International Aircraft Cabin Safety Symposium*, The Southern California Safety Institute.

Lazarus, R.S. and Folkman, S. (1984), *Stress, appraisal and coping*, Springer, N.Y.

Lebber, O. (1997), *The Crisis Manager*, Lawrence Erlbaum Associates, Inc., Mahwah.

Liss, C.M. (1994), 'Violence in the workplace,' in *Canadian Medical Journal (Ottawa)*, vol. 154, No. 4.

Loh, S. (2001), 'An Investigation of Airline Cabin safety: A Taxonomy of Disruptive Passenger Behaviour' in *Proceedings of the Eleventh Int'l Symposium on Aviation Psychology*, Columbus, OH.

Lord-Jones, K. and Miller, R. (1997), 'Violence in the Workplace – Crew Members', in *Proceedings of the 14th International Aircraft Cabin Safety Symposium*, Southern California Safety Institute, pp. 32-43.

Lucas, E. (1987), 'Psychological Aspects of Travel', *Travel Medicine International, vol. 2, pp. 99-104.*

Macleod, N. (1998), 'Pilots are from Mars, Cabin Crew from Venus,' *CAT, Civil Aviation Training*, vol. 9, issue 7, pp. 22-28.

MacNab, A.J., Vachon, J., Susak, L.E. and Pirie, G.E. (1900), 'In-flight stabilization of oxygen saturation by control of altitude for severe respiratory insufficiency', *Aviation, Space and Environmental Medicine*, vol. 61, pp. 829-832.

Marske, C.E. (1997), *'Safety as A Process'*, paper presented at the NTSB Symposium on Corporate Culture and Transportation Safety.

Marske C.E. (ed) (1991), Communities of Fate – Readings in Social Organization and Risk, University Press of America, Lanham.

Mattman, J.W. and Kaufer,S. (1995) *Complete workplace violence prevention manual*, Newport Beach, California, Workplace Violence Research Institute.

McKenzie-James, A., (in press) 'Air Rage' Handling Disruptive or Unruly Behaviour', in *Proceedings of the Seventeenth Annual International Aircraft Cabin Safety Symposium*, Southern California Safety Institute, 2000.

Meshkati, N. (1997), 'Human Performance, Organizational Factors and Safety Factors', paper presented at the NTSB Symposium on Corporate Culture and Transportation Safety.

Muir, H. and Cobbett, A. (1995), 'Flight Attendant Behaviour in Emergency Evacuations', in *Proceedings of the 12th International Aircraft Cabin Safety Symposium*, Southern California Safety Institute, pp. 108-119.

Nash, T. (1999), 'CAT talks to British Airways' General Manager Flight Training Tim Hodgson, *CAT, Civil Aviation Training*, pp. 40-43, June.

Parasuraman, A., Zeithaml, V. A. and Berry, L. L. (1986), 'SERVQUAL: A multiple-item Scale for Measuring Customer Perceptions of Service Quality,' Report No. 86-108, Marketing Science Institute, Cambridge, MA.

Parasuraman, A., Zeithaml, V. A. and Berry, L. L. (Fall 1991), 'Perceived Service Quality as a Customer-Based Performance Measure. An Empirical Examination of Organizational Barriers Using an Extended Service Quality Model', in *Human Resource Management*, 420-450.

Pathological Gambling: A Critical Review, National Academy Press, Washington, DC, 1999.

Paton, S. (1989), 'Medical studies in aviation: VII. Effects of low oxygen pressure on the personality of the aviator', *Aviation, Space and Environmental Medicine*, vol. 60, pp. 1225-6.

Peterson, J. (1997), 'Handling of Unruly Passengers, The Singapore Airline Experience', in *Proceedings of the 14th International Aircraft Cabin Safety Symposium*, Southern California Safety Institute, pp. 46-52.

Pidgen, N.F. and O'Leary, M. (1994), 'Organizational safety culture: Implications for aviation practice', in Johnston, N., McDonald. N., & Fuller, R. (Eds.), *Aviation Psychology in Practice*. Avebury Technical, Aldershot, pp. 21-43.

Phillips, S., Cabin Crew CRM, *The Journal of Professional Aviation Training*, vol. 1,

Number 6, pp. 17-21, July-August, 1999.

Pieren, A. (1997), 'The Occupational Health Factors of Flight Attendants', in *Proceedings of the 14ᵗʰ International Aircraft Cabin Safety Symposium,* Southern California Safety Institute, pp. 270-286.

Pizzino, A., Report on CUPE's (Canadian Union of Public Employees) National Health and Safety Survey of Aggression Against Staff, Ottawa, 1994 (January).

Polak, J., (1997), Sicherheit als Entscheidungsfaktor fur die Wahl der Fluggesellschaft. Thesis, Fachhochschule Munchen.

Policy and Guideline for Handling of Unruly Passengers (1998), internal publication, Swissair.

Poyner, B. and Warne, C. (1988), Tavistock Institute Of Human Relations: *Preventing violence to staff,* London, Health and Safety Executive, p. 7.

Prew, S., (1999) 'Unruly Passengers, the role of the police, *Aviation Security international,* pp. 24-27, February.

Prew, S. (1999), 'Training to combat air rage', *CAT Magazine, The Journal For Civil Aviation Training,* pp. 34-39, June.

Rayman, R. (1997), 'Passenger safety, health and comfort: a review', *Aviation Space and Environmental Medicine,* vol. 68, 5, pp. 432-440.

Reason, J. (1990), *Human Error.* Cambridge: Cambridge University Press, pp. 200-203, 208.

Reason, J. (1997), *'Corporate Culture and Safety',* paper presented at the NTSB Symposium on Corporate Culture and Transportation Safety.

Reice, H. and Beyer, J.M. (1993), *The cultures of work organizations,* Prentice Hall, New York.

Rosenbluth, Hal F. and McFerrin Peters, D. (1992), *The Customer Comes Second,* William Morrow and Company, Inc., New York.

Santangelo, E. (1997), 'Interference with Crewmembers in the Performance of Their Duties', in *Proceedings of the 14th Annual International Aircraft Cabin Safety Symposium and Technical Conference,* Southern California Safety Institute, pp. 76-79.

Scharfetter, C. (1991), *Allgemeine Psychopathology,* Thieme Verlag, Stuttgart, pp. 262.

Schein, E.H. (1992), *Organizanional culture and leadership,* Jossey-Bass, San Francisco.

Smith, P. (1986), *It Seems Like Only Yesterday,* McClellan and Stewart Limited, Toronto, p. 22.

Suggs, C.W. (1997), *'Corporate Culture and Transportation Safety',* paper presented at the NTSB Symposium on Corporate Culture and Transportation Safety.

Sundstrom, E. (1978), 'Crowding as a sequential process', in A. Baum and Y. M. Epstein (eds), *Human response in crowding,* Lawrence Erlbaum Associates, Hillsdale, New Jersey, pp. 32-116.

Suthichoti, S., (1995), 'Cultural Differences in the Perception of Safety: A Case Study', in *Proceedings of the 12th Annual International Aircraft Cabin Safety Symposium,* Southern California Safety Institute, pp. 80-84.

Tebo, P.V. (1997), *'Critical Elements of DuPont's Safety Culture',* paper presented at the NTSB Symposium on Corporate Culture and Transportation Safety.

Transport Canada, (1996), Human Factors For Aviation, Advanced Handbook. Ottawa.

Video Gambling In Foreign Air Transportation, Report to the Congress, Department of Transportation, March 1996.

Vogel, J.L. (1995), 'Relationship between aviation physiology and aviation psychology', in R. Fuller, N. Johnston and N. McDonald. (eds.), *Human Factors in Aviation Operations,* Avebury Technical, Aldershot, pp. 235-240.

Von Muelmann, M. (1995), 'Management of Sick and Invalid Passengers – First Aid Training for Cabin Attendants', in *Proceedings of the 12th Annual International Aircraft Cabin Safety Symposium,* Southern California Safety Institute, pp. 210-229.

Webb, J.T. and Pilmanis, A.A. (1993), Breathing 100% oxygen compared with 50% oxygen: 50% nitrogen reduces altitude-induced venous gas emboli, *Aviation, Space and Environmental Medicine,* Vol. 64, pp. 808-812.

Westrum, R. (1997), *'Safety of a Technological System ',* paper presented at the NTSB Symposium on Corporate Culture and Transportation Safety.

Westrum, R., (1995). Organizational dynamics and safety. In Johnston, N., McDonald. N., & Fuller, R. (eds.), *Applications of Psychology to the Aviation System.* Avebury Aviation, Aldershot.

Williams, G. (1993) *The Airline Industry and the Impact of Deregulation,* Ashgate Publishing: UK, p. 2.

Wood, R.H. (1995), 'Put the Passenger on the Team', in *Proceedings of the 12th Annual International Aircraft Cabin Safety Symposium,* Southern California Safety Institute, pp. 14-18.

Wood, R.H. (1991), Aviation Safety Programs, A Management Handbook, IAP, Inc., Casper.

Index